LANCASHIRE LIBRARY
BURNLEY

ZB 9/94

S. & T. L.

19. JUL. 1982
30. DEC. 1983
28. MAR. 1984
-9. MAY 1984
-7. JAN. 1985
22. AUG. 1985
14. JUL. 1986
27. MAY 1987
19. AUG. 1987
25. SEP. 1987
16 OCT 1987
12. FEB. 1988
17. FEB. 1990
16. AUG. 1990

WITHDRAWN FROM LANCASHIRE LIBRARIES

AUTHOR
NELSON, J. L.

CLASS
629.13435

TITLE
Lightplane engines.

No.
460783343

LANCASHIRE COUNTY COUNCIL

This book should be returned on or before the latest date shown above to the library from which it was borrowed

LIBRARY HEADQUARTERS, 143 CORPORATION STREET, PRESTON, PR1 8RH

Lightplane Engines —2nd Edition

Other TAB books by the author:

No. 2228 *Practical Guide to Aviation Weather*
No. 2259 *Complete Guide to Radio Navigation for Private & Commercial Pilots*
No. 2295 *Pilot's Digest of FAA Regulations–New Revised 2nd Edition*

Lightplane Engines –2nd Edition

by John L. Nelson

MODERN AVIATION SERIES

TAB BOOKS Inc.

BLUE RIDGE SUMMIT, PA. 17214

LANCASHIRE LIBRARY

02150967

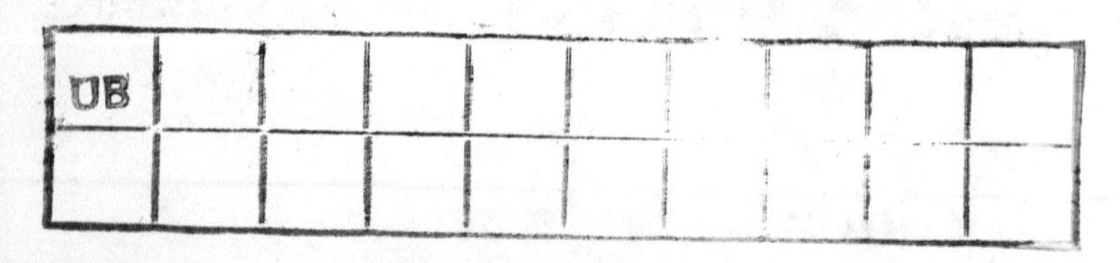

46078334 3

SECOND EDITION

FIRST PRINTING

FEBRUARY 1981

Copyright © 1981 by TAB BOOKS Inc.

Printed in the United States of America

Reproduction or publication of the content in any manner, without express permission of the pblisher, is prohibited. No liability is assumed with respect to the use of the information herein.

Library of Congress Cataloging in Publicaion Data

Nelson, John Lewis, 1926-
Lightplane engines.

Published in 1974 under title: Modern lightplane engines.
Includes index.
1. Airplanes—Motors. I. Title.
TL701.N4 1981 629.134'352 80-28161
ISBN 0-8306-2323-X (pbk.)

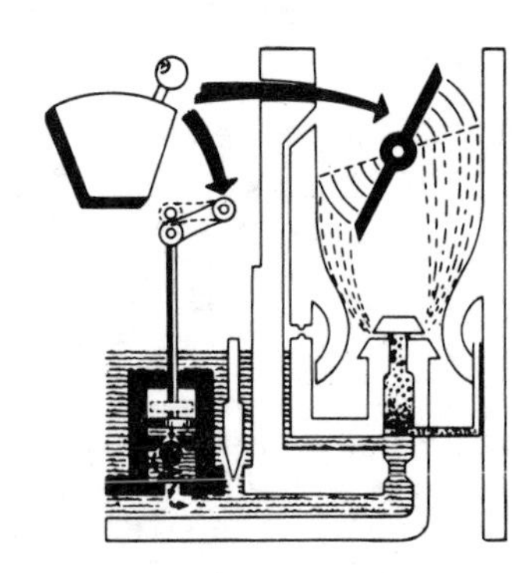

Contents

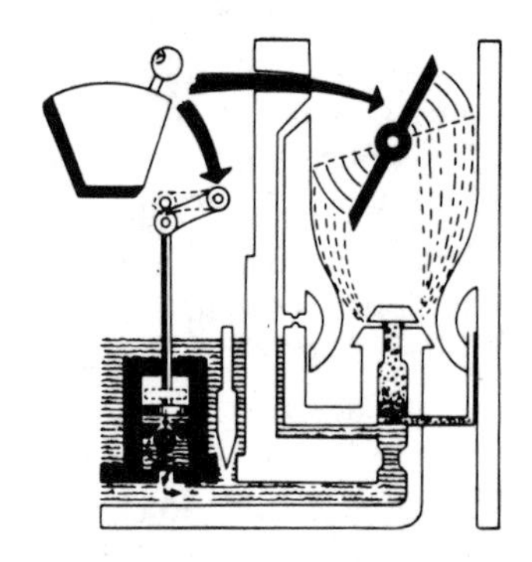

Preface

This book is for lightplane pilots who want to become wiser and more efficient (and perhaps *older*) flyers by learning a bit about engines. It is also for the amateur mechanic, the homebuilt-plane enthusiast and anyone who, out of preference or necessity, performs engine maintenance.

Of course, federal aviation regulations require that major repairs done by the amateur be "signed off" by a licensed airplane engine mechanic. This means that such work must be done under direct supervision and/or inspected and approved by a properly certificated mechanic. And I should also stress that this book does not presume to replace the manufacturer's manuals for the engines discussed herein.

The author sincerely thanks the many people who made this text possible. To my wife Dolores and daughter Sandy my sincere appreciation for the many hours of typing and to Barbara Nelson my gratitude for the fine artwork. To Tom Cook, my appreciation for reviewing the text and the many fine technical comments. To editor Joe Christy, my thanks for the editing of *Lightplane Engines*. And to the FAA and the many manufacturers that contributed technical literature—thank you.

Technical Data Contributors

A.C. Sparkplug Division, General Motors, Flint, Michigan
Aircraft Owners and Pilots Association, Washington, D.C.
Alcor Aviation Inc., San Antonio, Texas

Avco Lycoming, Williamsport, Pennsylvania
Bede Aircraft Inc., Cleveland, Ohio
Champion Sparkplug Co., Toledo, Ohio
Chevron Research Co., Richmond, California
Chrome Plate Inc., San Antonio, Texas
Federal Aviation Administration, Washington, D.C.
H.P. Books, Tucson, Arizona
Marvel-Schebler/Tillotson, Decatur, Illinois
National Transportation & Safety Board, Washington, D.C.
Phillips Petroleum Co., Bartlesville, Oklahoma
Prestolite Division, Eltra Corp., Toledo, Ohio
Rajay Industries Inc., Long Beach, California
Richter Aero Equipment, Essex, New York
Slick Electro Inc., Rockford, Illinois
Teledyne Continental Motors, Mobile, Alabama
T.W. Smith Aircraft Inc., Cincinnati, Ohio

John L. Nelson

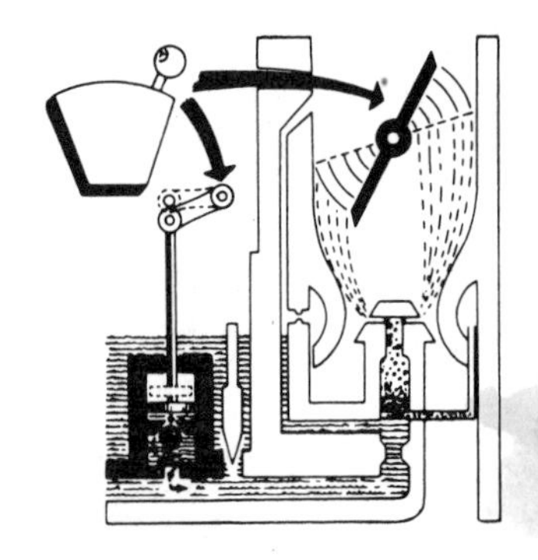

About The Author

John L. Nelson's interest in aviation began in the mid-1930s as an early member of the Academy of Model Aeronautics. Many trips were made to the local airport to test fly rubber and gas powered models, watch genuine aircraft, and stare at the zeppelins Akron and Macon. Then came World War II and service with the USAF including aviation cadet training. With the war's end in 1945, college followed and a degree in electrical engineering was earned. The aerospace age dawned in 1950 and with it many new opportunities. Research and development in the field of airborne radar and cathode ray tube devices has filled the years since then. As a certificated flight and ground school instructor the author has enjoyed teaching for eight years.

Introduction

Powered flight began with lighter-than-air machines. On September 24, 1852, Henri Giffard's 144-foot dirigible flew some 17 miles from Paris to Trappes. It was fitted with a 3-horsepower steam engine that weighed 350 pounds (117 lbs. per hp!). This event, man's first powered flight, established aviation's initial speed record as 5 mph.

More than half a century passed before Orville and Wilbur Wright designed, built and demonstrated the first successful airplane engine. The Wright engine weighed 156 pounds and produced 12-horsepower (13 lbs per hp). Compared to other gasoline engines in 1903—which averaged 80 or more pounds per horsepower—it represented an achievement almost as great as the flying machine it propelled.

In spite of—or perhaps because of—seven decades of experimenation and development, the reciprocating engines that power modern lightplanes operate on the same principle as the one that flew the world's first successful airplane in 1903. These powerplants employ the four stroke principle—*intake, compression, power,* and *exhaust* strokes—which is (technically) the *Otto Cycle* (named for Nickolas Otto who perfected it in 1876).

Otto Cycle

In these engines, the crankshaft revolves twice for each complete cycle of four strokes, making two outward strokes (away from the crankshaft) and two inward strokes (toward the crankshaft). Therefore, each cylinder "works" every other revolution of the crankshaft, while the piston coasts through on the intervening revolution and forces out the burned gases produced by the working stroke of the previous revolution.

A complete cycle (Fig. 1-1) goes like this: As the intake stroke begins, the piston is near top dead center (TDC) and the intake valve is opened by its camshaft lobe. The piston travels inward. This sucks the fuel/air mixture into the cylinder from the intake manifold. When the piston reaches the end of its intake stroke, the intake valve closes and the compression stroke begins as the piston now moves outward toward the top of the cylinder. A split-second before the piston reaches TDC, the magnetos send an electrical impulse to the spark plug. This ignites the compressed fuel/air mixture.

While this controlled explosion takes place, the piston passes TDC and, propelled by this explosion, moves inward on its power or working stroke. Then, when the piston reaches the bottom of the cylinder, the exhaust valve opens and the piston coasting through on the exhaust stroke purges the cylinder of the burned gases. Now, the intake valve opens again and a new cycle begins. Your automobile engine (unless you have a Wankle-powered car) operates on this same principle.

Otto Cycle engines are only about one-third thermally efficient. That is, the crankshaft receives but one-third of the energy released by the burning fuel/air mixture. About one-tenth of the total is lost to engine friction, a similar amount to cooling.

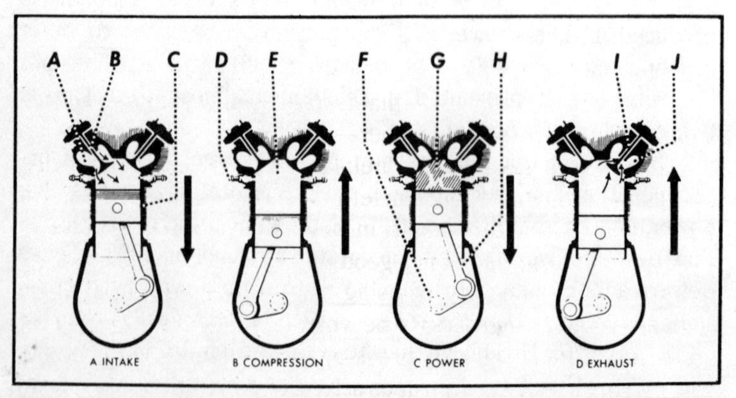

Fig. 1-1. Otto Cycle: A. fuel mixture being drawn into the cylider; B. intake valve; C. piston; D. spark plug; E. fuel mixture being compressed; F. crankshaft; G. fuel mixture after ignition; H. connecting rod; I. exhaust valve; J. burned gases being expelled from the combustion chamber.

Almost half of the energy released in the combustion chamber is lost (in the form of unused heat and pressure) through the exhaust ports.

Therefore, from an engineering standpoint, the modern lightplane engine is very inefficient. It is, however, the best compromise so far offered (Fig. (1-2). We say "compromise" because factors other than thermal efficiency are equally important: cost of operation, weight, reliability, ease of maintenance and fuel tankage needed.

The valves are actuated by a camshaft. Tearshaped lobes on the camshaft are so arranged as to open each valve at exactly the right instant. The camshaft is geared to revolve at one-half crankshaft speed because the intake and exhaust valves open and close only once for each two revolutions of the crankshaft.

Valve timing differs slightly with different engines. On many supercharged powerplants, the intake valve opens before the exhaust valve is completely closed. The result is better expulsion of the burned gases. This is called *valve overlap*.

The spark from the ignition system is nearly always supplied to the combustion chamber before the piston reaches TDC. This is called, *spark advance*. In other words, a 15-degree spark advance means that "fire" is supplied when the piston is still 15-degrees below TDC and before completion of the compression stroke.

Compression ratio is the ratio of the cylinder volume (expressed in cubic inches) at the end of the intake stroke to the cylinder volume at the end of the compression stroke. It is, in short, a direct measure of how much the fuel/air mixture is squeezed before ignition.

Fuel

The ratio of fuel-to-air—that is, the relative amounts of each entering the combusion chamber—greatly affects engine operation. A very lean mixture (less fuel; more air), can cause backfiring or detonation due to incomplete or premature ignition. An excessively rich mixture results in loss of power and waste of fuel.

Speaking of fuel, this is a good time to explain octane ratings and stress the necessity of using airplane gasoline in airplane engines. Aviation fuels must be volatile. They must evaporate easily and therefore quickly break-up into fine drops in the intake manifold to be ready to burn upon arrival of the ignition spark; from carburetor to cylinder, it takes about one-sixteenth of a second.

Also, avgas must resist knocking within the pressure ranges established by the compression ratio of the engine in which it is

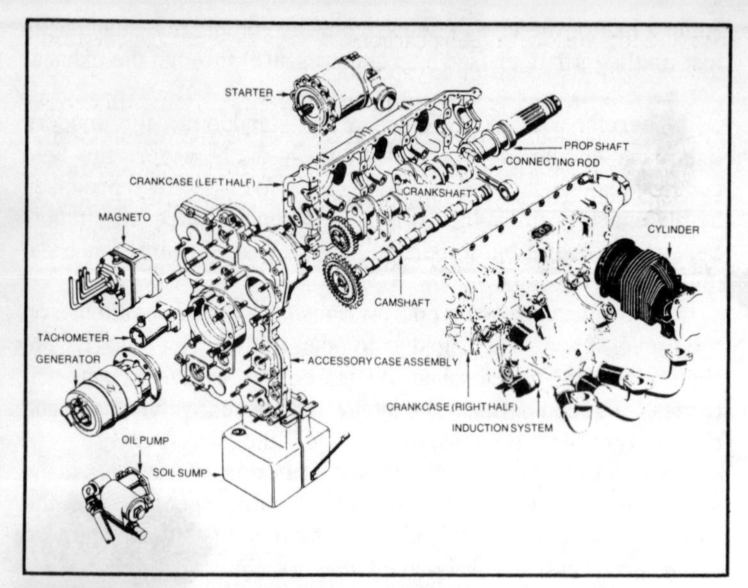

Fig. 1-2. An exploded view of a typical six cylinder opposed engine (courtesy of FAA).

used. Knock value is given in octane ratings because it is determined by comparing the knock value of a given fuel with the knock value of a reference fuel mixture of isooctane and heptane. When there is a double rating, such as 80/87 or 100/130, the first number is the octane rating for a lean mixture and the second the octane rating for a rich mixture. Numbers below 100 are octane numbers, while numbers above 100 are performance numbers.

Other important properties of airplane gasoline include a high boiling point to prevent vapor locking (avgas vaporizes at about 140 degrees F at sea level; 100 degrees F at 20,000 ft.). This is not the same things as volatility. The goal in avgas blending is to produce a fuel that has maximum resistance to boiling, yet evaporates easily.

Also important is a fuel that is stable and retains all its properties in storage. It must not corrode or otherwise cause deterioration of the metals and other materials in the plane's engine and fuel system. You cannot safely use an automotive motor fuel in your light airplane. The molecular structure of gasoline is different. True, both might have an octane rating of, say, 90; but octane rating is only part of the story.

Aviation gasoline is largely composed of high-heat energy paraffins in order to achieve the previously described properties. Automotive gasoline is made up of large quantities of comparatively unstable olefin hydrocarbons. Olefins cause gum deposits and are highly susceptible to vapor lock.

Dye is added to avgas in small amounts to allow easy identification of the three grades at the pumps and in your tanks: 80/87—red; 100/130—green; 115/145—purple. You probably have the habit (its a good one) of *visually* checking your fuel tanks after each fuel purchase. Also, *smell* them. More than one lightplane has been lost after taking on jet fuel by mistake.

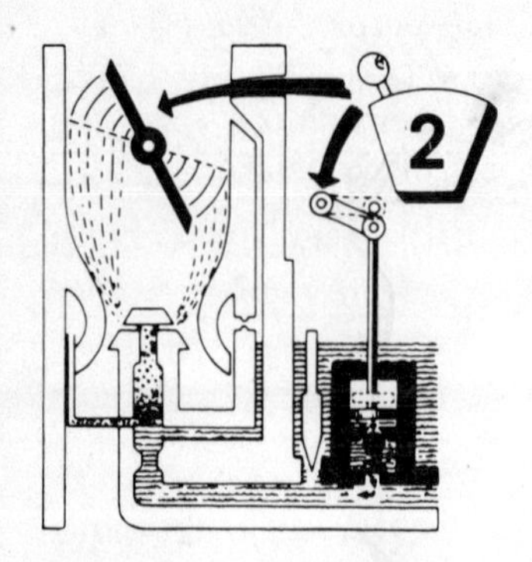

Engine Construction: General

All production lightplanes today are fitted with opposed type engines. This means that the cylinders, in two banks, are horizontally positioned on opposite sides of the crankcase. Such engines will operate just as well mounted vertically (which is the case with some helicopters), but of course the low silhouette and easier streamlining—especially when wingmounted—makes horizontal installations the rule.

The crankshaft transforms the reciprocating motion of the pistons and connecting rods into rotary motion for rotation of the propeller. The connecting rods attach to crank pins in the off-set portions of the crankshaft. These off-sets constitute the "crank" part of the crankshaft. The crankpin is usually hollow. This cuts weight and provides an oil passage for lubrication.

Excessive vibration in an engine promotes fatigue failure and rapid wear to moving parts. It can be caused by a crankshaft which is not balanced. Crankshafts are balanced for static and dynamic conditions. Static balance is tested by placing the crankshaft on two knife edges. If the shaft tends to turn toward any one position, it is out of static balance. A crankshaft is dynamically balanced when little or no vibration is produced during engine operation. To reduce vibration to a minimum during engine operation, dampers are attached to the crankshaft counterweights. These dampers are small pendulums that are limited in frequency and vibrating movement to correspond to the frequency of the engine's power

LANCASHIRE LIBRARY

impulses. When the vibration frequency of the crankshaft occurs, the pendulums oscillate out-of-time with the crankshaft vibration—therefore countering it (Fig. 2-1).

Pistons

The majority of aircraft engine pistons are machined from aluminum alloy forgings. Grooves are machines in the outside surface to receive the piston rings and cooling fins are provided on the inside of the piston to aid heat transfer to the engine oil. The top face of the piston (or head) can be flat, convex, or concave. Recesses can be machined in the piston head to prevent interference with the valves.

As many as four grooves can be machined around a piston to accommodate compression rings and oil rings. Compression rings are installed in the uppermost grooves; oil control rings are typically installed immediately above and below the piston pin. The piston is usually drilled at the oil control ring grooves to allow surplus oil scraped from the cylinder walls to pass back into the crankcase. The oil control ring installed at the base of the piston wall or skirt prevent excessive oil consumption.

Gray cast iron is most often used in making piston rings. In some engines, chrome-plated mild steel rings are used in the top compression ring groove because such metal can better withstand the high temperatures present at this point.

Oil control rings regulate the thickness of the oil film on the cylinder wall. If too much oil enters the combustion chamber, it will burn and leave a thick coating of carbon on the combustion chamber walls, piston head, spark plugs and valve heads. This carbon can cause the valves and piston rings to stick if it enters the ring grooves and valve guides. It can also cause spark plug misfiring, preignition and excessive oil consumption.

Cylinders

The cylinder head of an air-cooled engine is usually made of an aluminum alloy to gain a low weight-to-strength ratio, along with good heat transfer properties. Cylinder heads are forged or die-cast for greater strength. The inner shape of a cylinder head can be flat, semispherical or peaked like a roof. The semispherical type has proved most satisfactory because it is stronger and aids in a more rapid and thorough purging of the exhaust gases.

A cylinder had two major parts: the cylinder head and the cylinder barrel (Fig. 2-2). At assembly, the head is expanded by heating and then screwed down on the cylinder barrel which has

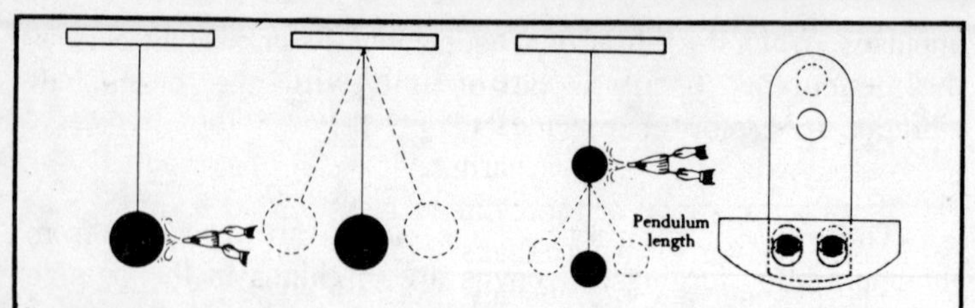

If a simple pendulum is given a series of regular impulses at a speed corresponding to its natural frequency (using a bellows to simulate a power impulse in an engine) it will commence swinging, or vibrating, back and forth from the impulses. Another pendulum, suspended from the first, would absorb the impulses and swing itself, leaving the first stationary. The dynamic damper is a short pendulum hung on the crankshaft and tuned to the frequency of the power impulses to absorb vibration in the same manner.

Fig. 2-1. Basic principles of a dynamic damper (courtesy of FAA).

been chilled. Then when the head cools and contracts and the barrel warms and expands a gas-tight joint results.

After casting, the spark plug bushings, valve guides, rocker arm bushings and valve seats are installed in the cylinder head. Spark plug holes are fitted with bronze or steel bushings that are shrunk and screwed into the openings. Stainless steel Heli-Coil

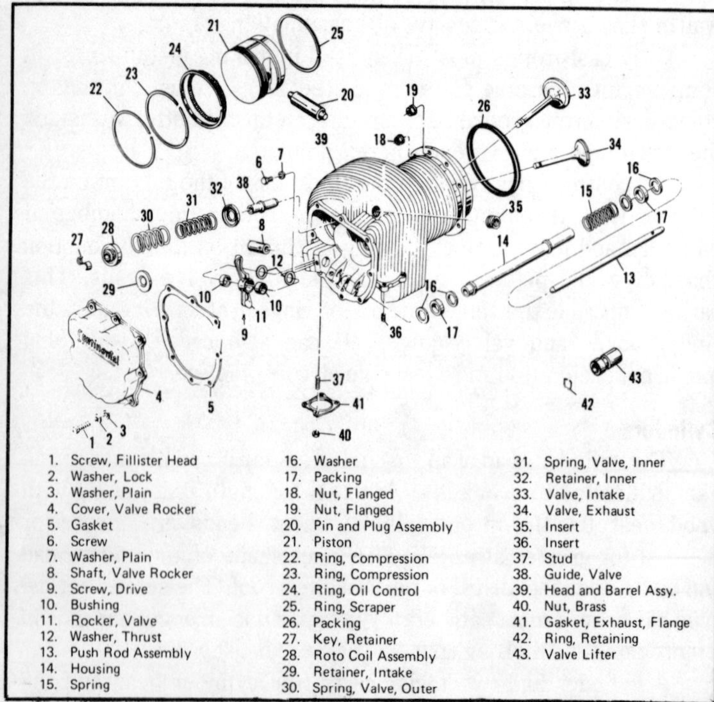

Fig. 2-2. Component parts of a cylinder as employed in the Continental 10-520 engine (courtesy of Teledyne Continental Motors).

spark plug inserts are used in some engines. Bronze or steel valve seats are usually shrunk or screwed into drilled openings in the cylinder head to provide guides for the valve stems. The valve seats are circular rings of hardened metal that protect the relatively softer metal of the cylinder head from the hammering action of the valves and from exhaust gas erosion.

The cylinder barrel is usually made of a steel alloy forging with the inner surface hardened to resist wear. This hardening can be accomplished by exposing the steel to ammonia or cyanide gas while the steel is very hot. The steel soaks up nitrogen from the gas which forms iron nitrides on the exposed surface. The cylinder barrel is then said to be *nitrided*.

The propeller shaft end of the engine is always the front end, and the accessory end is the rear end, regardless of how the engine is mounted in an aircraft. When referring to the right or left side of an engine, always assume you are viewing it from the rear or accessory end. However, the numbering of cylinders in an opposed-type engine is by no means standard. Some are numbered from the rear and others are numbered from the front by the manufacturers. It's necessary to check the appropriate engine manual to determine the cylinder numbering system of a given engine.

The firing order of the cylinders (the sequence in which combusion takes place in each) is a function of dynamic balance. Firing order in opposed engines can usually be listed in pairs of cylinders. The firing order of six-cylinder opposed engines is 1-4-5-2-3-6. The firing order of one model four-cylinder opposed engine is 1-4-2-3; but on another it is 1-3-2-4.

Valves

Intake valves operate at lower temperatures than exhaust valves and are therefore made of chrome-nickel steel. Exhaust valves are usually made of nichrome, silchrome or cobalt-chromium steel.

The valve head (Fig. 2-3) has a ground face which forms a seal against the valve seat in the cylinder head when the valve is closed. The face of the valve is usually ground to an angle of either 30 or 45 degrees. In some engines, the intake valve face is ground to an angle of 30 degrees while the exhaust valve face is beveled to 45 degrees (Fig. 2-4).

The valve lift (distance the valve is lifted off its seat) and the valve duration (length of time valve is held open) are both determined by the shape of the camshaft lobes and the rocker arm

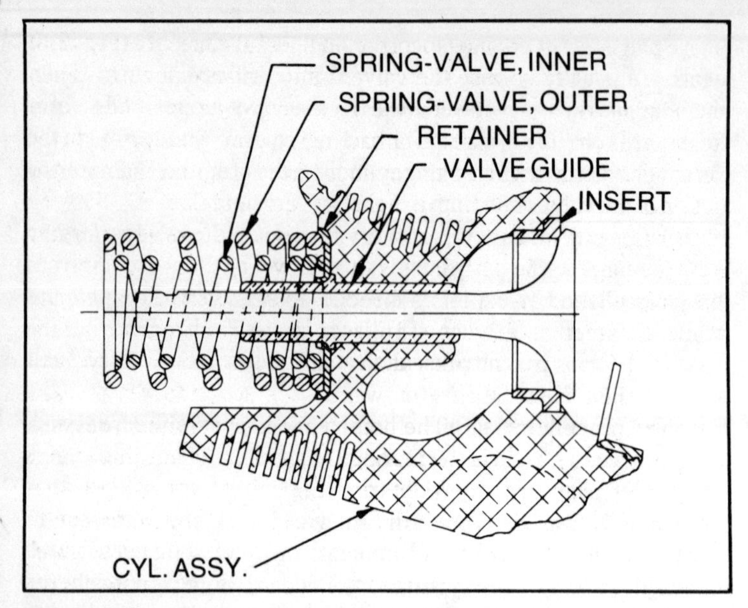

Fig. 2-3. Valve detail of the Continental 10-520 engine (courtesy of Teledyne Continental Motors).

lever distances. The camshaft, rotating at one-half crankshaft speed, lifts the tappets with these lobes which in turn forces the push rods and rocker arms to open the valves. A hole is drilled through each tappet to allow oil to flow to the hollow push rods and the rocker assemblies.

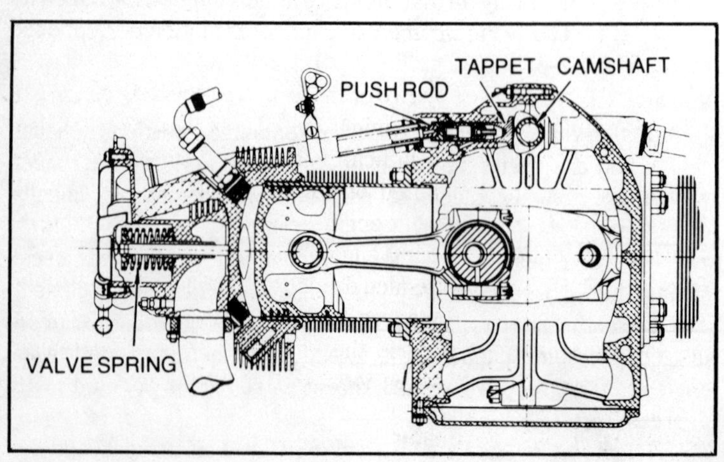

Fig. 2-4. Cross section view of engine cylinder and valve operating mechanism (courtesy of FAA).

Some aircraft engines incorporate hydraulic tappets (Fig. 2-5) which automatically keep the valve train clearance at zero. When the engine valve is closed, the face of the tappet body (cam follower) is on the base circle or back of the cam. The light plunger spring lifts the hydraulic plunger so that its outer end contacts the push rod socket, exerting a light pressure against it, thereby eliminating any clearance in the valve linkage. As the plunger moves outward, the ball check valve moves off its seat. Oil from the supply chamber, which is directly connected with the engine lubrication system, flows in and fills the pressure chamber.

As the camshaft rotates, the cam pushes the tappet body and the hydraulic lifter cylinder outward. This action forces the ball check valve onto its seat. The body of oil trapped in the pressure chamber acts as a cushion. During the interval when the engine valve is off its seat, a predetermined leakage occurs between plunger and cylinder bore which compensates for any expansion or contraction in the valve train. Immediately after the engine valve closes, the amount of oil required to fill the pressure chamber flows in from the supply chamber to prepare for another cycle of operation.

The rocker arms transmit the lifting force from the cams to the valves. Generally, one end of the arm bears against the push rod and the other bears on the valve stem. The arm can have an adjusting screw for adjusting the clearance between the rocker arm and the valve stem tip. The screw is adjusted to the specified clearance to make certain that the valve closes fully. No adjustment provision is necessary in some valve systems that enjoy hydraulic valve lifters.

Each valve is closed by two or three helical-coiled springs. If a single spring were used, it might vibrate or surge at certain speeds. To eliminate this difficulty, two or more springs (one inside the other) are installed on each valve. Each spring will therefore vibrate at a different engine speed and rapid damping of spring-surge vibrations during engine operation is accomplished. Two or more springs also reduce danger of weakness and possible failure by breakage due to heat and metal fatigue. The functions of the valve springs are to close the valve and to hold the valve securely on the valve seat when not actuated by the cam.

Exhaust Manifold Cabin Heater

Concentrations of carbon monoxide exceeding one part in 20,000 parts of air are hazardous. To prevent an airplane from becoming a deathtrap, a thorough examination of the exhaust manifold and heater assembly should be conducted at regular intervals or whenever carbon monoxide contamination of the cockpit or cabin is suspected. Cracks and holes can develop in an exhaust manifold in a relatively short time. Some aircraft manufacturers recommend that exhaust and heater systems be inspected as often as every 25 hours of flight time. Carbon monoxide in the cabin or cockpit has been traced to worn or defective exhaust stack slip joints, exhaust system cracks or holes, openings in an engine firewall, "blowby" at the engine breather, defective gaskets in the exhaust manifold, defective mufflers and inadequate sealing or fairing around struts and fittings on an aircraft fuselage.

Brake Horsepower

It is "what's up front that counts" and so it is with brake horsepower (Fig. 2-6). Brake horsepower is simple indicated horsepower minus losses due to friction and engine accessories. In modern aircraft engines, the power loss due to friction is typically 10 percent of the indicated horsepower. Engine accessories could add an additional loss of 1 to 2 percent.

The purpose of a propeller is to convert engine brake horsepower into thrust. This it does at a typical efficiency of 80 to 85 percent for a well matched engine-propeller combination. The result is often termed *thrust horsepower*. A small, light airplane engine such as the Continental 0-200 can produce 114 indicated hp., 100 brake hp., and 80 thrust hp. when operated at sea level on standard day.

Engine Designation Code

A simple system of coding is used to described the many kinds of aircraft engines that exist today. Basically, the code denotes the

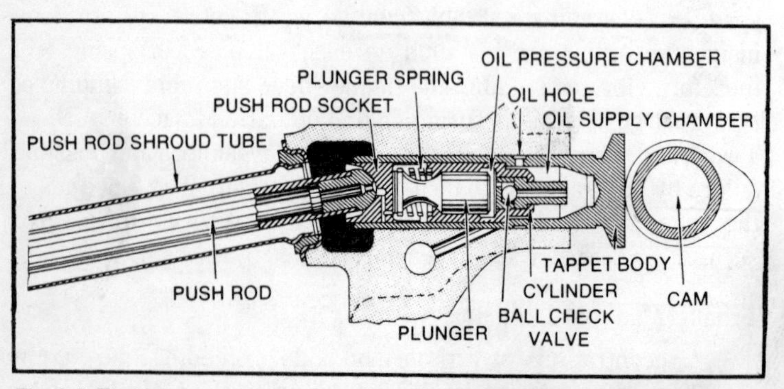

Fig. 2-5. Typical construction of a hydraulic tappet (courtesy of FAA).

Fig. 2-6. The Continental 10-520D engine. Rated 300hp at 2850RPM, the engine displaces 520 cubic inches and weighs only 456 pounds (courtesy of Teledyne Continental Motors).

type of engine, its displacement (cubic inches), and model as follows:

Type	Displacement	Model
TS10	520	B

The following abbreviations are used in the type code:

A - Aerobatic
G - Geared
H - Helicopter
I - Fuel Injected
L - Left Hand Rotation Crankshaft
M - Drone
O - Opposed Cylinders
S - Supercharged
T - Turbo-supercharged
V - Vertical

The digit "1" in the displacement designation (T10-541) indicates the model has an integral accessory housing (Lycoming).

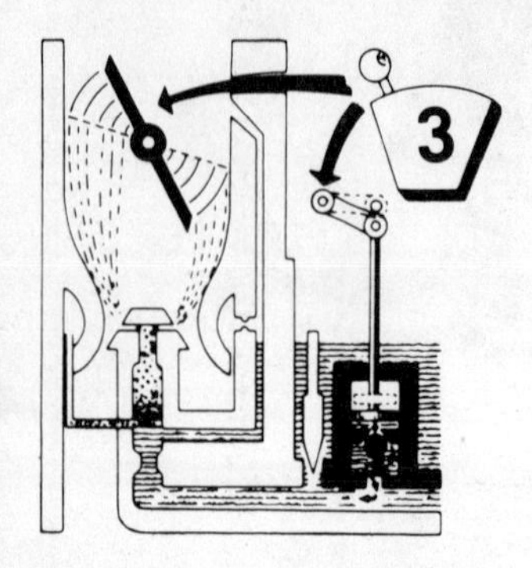

Carburetor Systems

Carburetors are almost as old as the internal combustion engine itself. The two basic types of carburetors employed by lightplane engines are the float type and the pressure type. Of these, the float type is by far the more common. Primary disadvantages of the float type are its susceptibility to ice and its limited performance during aerobatic flight.

Float-Type Carburetors

A float-type carburetor (Fig. 3-1) consists basically of a main air passage through which the engine draws its supply of air, a mechanism to control the quantity of fuel discharged in relation to the flow of air and a means of regulating the quantity of fuel-air mixture delivered to the engine cylinders.

The essential parts of float-type carburetor are the:

—Float mechanism and its chamber.
—Main metering system.
—Idling system.
—Mixture control system.
—Accelerating system.
—Economizer system.

A float chamber is located between the fuel supply and the metering system of the carburetor. The float chamber provides a nearly constant level of fuel to the main discharge nozzle. The fuel level must be maintained slightly below the discharge nozzle outlet holes to prevent fuel leakage from the nozzle when the engine is

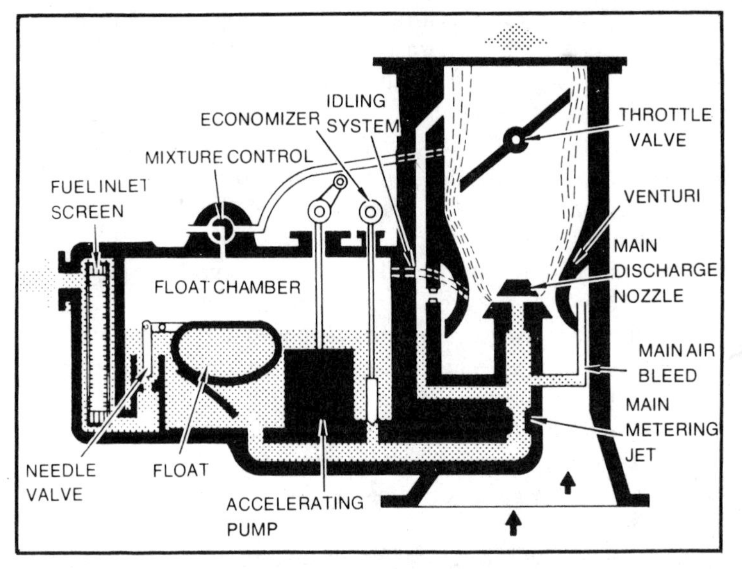

Fig. 3-1. Diagram of a basic float-type carburetor (courtesy of FAA).

not operating. The level of fuel in the flat chamber is kept nearly constant by means of a float-operated needle valve and a seat. With no fuel in the float chamber, the float drops toward the bottom of the chamber and allows the needle valve to open wide. As fuel is admitted from the supply line, the float rises and closes the valve at a predetermined level. When the engine is running and fuel is being drawn out of the float chamber, the valve assumes an intermediate position so that the valve opening is just sufficient to supply the required amount of fuel and keep the level constant. With the fuel at the correct level, the discharge rate is controlled accurately by the air velocity through the carburetor and the atmospheric pressure on top of the fuel in the float chamber. A vent or small opening in the top of the float chamber allows air to enter or leave the chamber as the level of fuel rises or falls. This vent passage is open into the engine air intake. Therefore, the air pressure in the chamber is always the same as that existing in the air intake.

The main metering system supplies fuel to the engine at all speeds above idling and consists of:

—A venturi.
—A main metering jet.
—A main discharge nozzle.
—A passage leading to the idling system.
—The throttle valve.

The venturi performs three functions: proportions the fuel-air mixture, decreases, the pressure at the discharge nozzle, and limits the airflow at full throttle. The fuel discharge nozzle is located in the carburetor barrel so that its open end is in the throat or narrowest part of the venturi. A main metering orifice, or jet, is placed in the fuel passage between the float chamber and the discharge nozzle to limit the fuel flow when the throttle valve is wide open.

As the air flows through the venturi, its velocity increases. This velocity increase creates a low pressure area in the venturi throat. The fuel discharge nozzle is exposed to this low pressure. Because the float chamber is vented to atmospheric pressure, a pressure drop across the discharge nozzle is created. It is this pressure difference, or metering force, that causes fuel to flow from the discharge nozzle. The fuel comes out of the nozzle in a fine spray and the tiny particles of fuel in the spray quickly vaporize in the air.

At low engine speeds, where the metering force is considerably reduced, the fuel delivery from the discharge nozzle decreases if an air bleed (air metering jet) is not incorporated in the carburetor.

In a carburetor (Fig. 3-2), a small air bleed is led into the fuel nozzle slightly below the fuel level. The open end of the air bleed is in the space behind the venturi wall where the air is relatively motionless and approximately at atmospheric pressure. The low pressure at the tip of the nozzle not only draws fuel from the float chamber but also draws air from behind the venturi. Air bled into the main metering fuel system decreases the fuel density and destroys surface tension. This results in better vaporization and control of fuel discharge. This is especially true at lower engine speeds.

The throttle, or butterfly valve, is located in the carburetor barrel near one end of the venturi. It provides a means of controlling engine speed or power output by regulating the airflow to the engine. With the throttle valve closed at idling speeds, air velocity through the venturi is so low that it cannot draw enough fuel from the main discharge nozzle; in fact, the spray of fuel might stop altogether.

However, low pressure (piston suction) exists on the engine side of the throttle valve. In order to allow the engine to idle, a fuel passageway is incorporated to discharge fuel from an opening in the low pressure area near the edge of the throttle valve. This opening is called the idling jet. With the throttle open enough so that the main discharge nozzle is operating, fuel does not flow out of the idling jet. As soon as the throttle is closed far enough to stop the spray from the main discharge nozzle, fuel flows out of the idling jet. A separate air bleed, known as the *idle air bleed,* is included as part of the idling system. It functions the same as the *main air bleed.* An idle mixture adjusting device is also incorporated.

As altitude increases, air becomes less dense. At an altitude of 18,000 feet, the air is only half as dense as it is at sea level. An engine cylinder full of air at 18,000 feet contains only half as much oxygen compared to a cylinder full of air at sea level. The low pressure area created by the venturi is dependent upon air velocity rather than air density. The action of the venturi draws the same volume of fuel through the discharge nozzle at a high altitude as it does at low altitudes. Therefore, the fuel mixture becomes richer as altitude increases. This can be overcome either by a manual or an automatic mixture control.

On float-type carburetors, two types of purely manual or cockpit controllable devices are in general use for controlling fuel-air mixtures: the needle type and the back-suction type. With the needle-type system, manual control is provided by a needle valve in the base of the float chamber. The back-suction type mixture control system is the most widely used. In this system, a certain amount of venturi low pressure acts upon the fuel in the float chamber so that it opposes the low pressure existing at the main discharge nozzle. An atmospheric line, incorporating an adjustable valve, opens into the float chamber. When the valve is completely closed, pressures on the fuel in the float chamber and at the discharge nozzle are almost equal and fuel flow is reduced to maximum lean. With the valve wide open, pressure on the fuel in the float chamber is greatest and fuel mixture is richest. Adjusting the valve to positions between these two extremes controls the mixture.

When the throttle valve is opened quickly, a large volume of air rushes through the air passage of the carburetor. However, the amount of fuel is that is mixed with the air is less than normal. This is because of the slow response rate of the main metering system. As a result, after a quick opening of the throttle, the fuel-air mixture leans out momentarily. To overcome this tendency, a carburetor is equipped with a small fuel pump called an accelerating pump. It consists of a simple piston pump operated through linkage, by the throttle control, and a line opening into the main

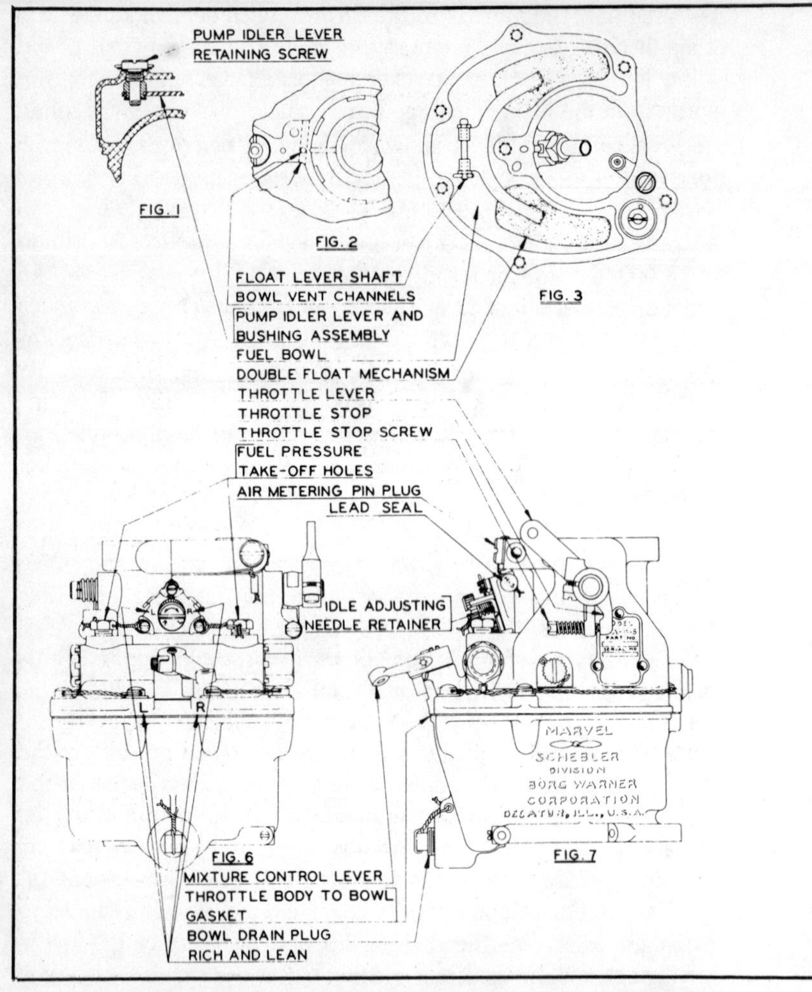

Fig. 3-2. The Marvel-Schebler model MA4-5 carburetor. A popular device, the MA4-5 is employed in a number of the high-horsepower output Lycoming and Continental engines (courtesy of Marvel-Schebler).

metering system or the carburetor barrel near the venturi. When the throttle is closed, the piston moves back and fuel fills the cylinder. If the piston is pushed forward slowly, the fuel seeps past it back into the float chamber. But if it is pushed rapidly, it will emit a charge of fuel and enrich the mixture in the venturi.

For an engine to develop maximum power at full throttle, the fuel mixture must be richer than for cruise. The additional fuel is used for cooling the engine to prevent detonation. An *economizer* is

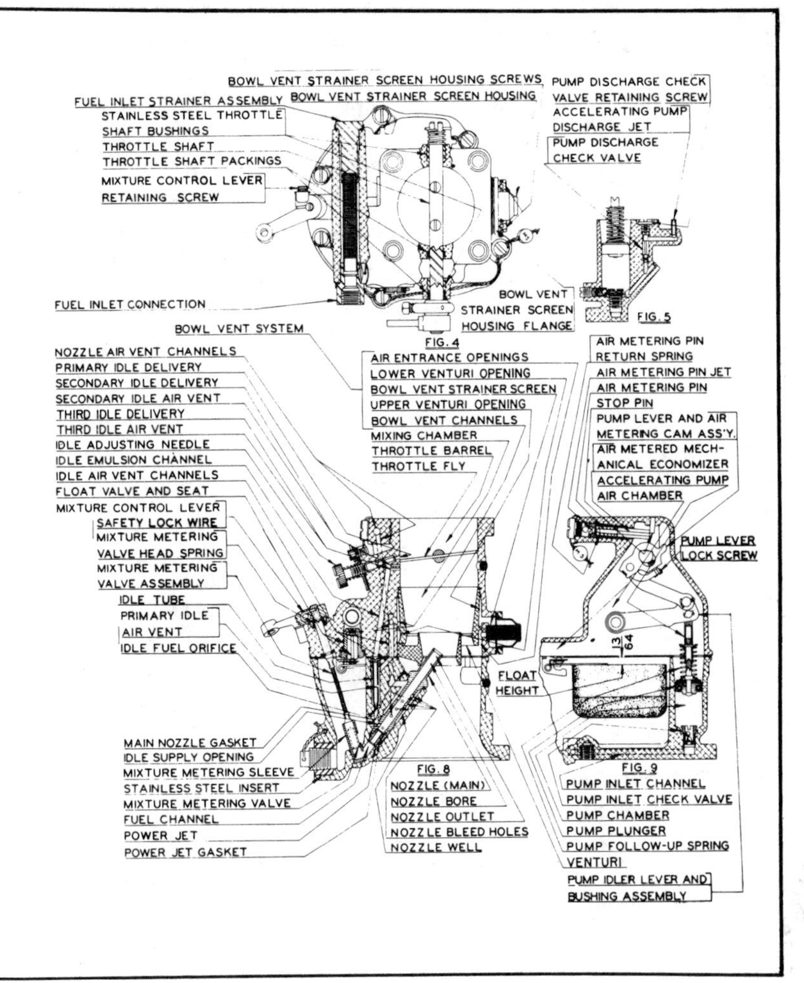

essentially a value which is closed at throttle settings below approximately 60 to 70 percent of rated power. This system, like the accelerating system, is operated by the trottle control. A typical economizer system consists of a needle valve which begins to open when the throttle valve reaches a predetermined point near the wide-open position. As the throttle continues to open, the needle valve is opened further and additional fuel flows through it.

Pressure Injection Carburetors

Pressure injection carburetors are distinctly differently from float-type carburetors. They do not incorporate a vented float chamber or suction pickup from a discharge nozzle located in the venturi tube. Instead, they provide a pressurized fuel system that is closed from the engine fuel pump to the discharge nozzle. The venturi serves only to create pressure differentials for controlling the quantity of fuel to the metering jet in proportion to airflow to the engine.

The injection carburetor is a hydromechanical device that meters fuel through fixed jets according to the mass airflow through the throttle body and discharges it under a positive pressure. The pressure injection carburetor is an assembly of the following units:

—Throttle body.
—Regulator unit.
—Fuel control unit.

The throttle body contains the throttle valves, the main venturi and the boost venturi. All air entering the cylinders must flow through the throttle body; therefore, it is the air control and measuring device. The regulator unit serves to regulate the fuel pressure to the inlet side of the metering jets in the fuel control unit. This pressure is automatically regulated according to the mass airflow to the engine. The fuel control unit is attached to the regulator assembly and contains all metering jets and valves. The idle and power enrichment valves, together with the mixture control plates, select the jet combinations for the various power settings.

Direct Fuel-Injection Systems

The direct fuel-injection system is a second technique for mixing fuel and air. This technique is common on engines of higher horsepower and has certain advantages over a conventional carburetor system. There is less danger of induction system icing because the drop in temperature due to fuel vaporization takes place in or near the cylinder. Acceleration is also improved because of the positive action of the injection system. In addition, direct fuel injection improves fuel distribution. This reduces the over-heating of individual cylinders often caused by variation in mixture due to uneven distribution. The fuel injection system also gives better fuel economy than a system in which the mixture to most cylinders must be richer than necessary so that the cylinder with the leanest mixture will operate properly.

Continental Fuel Injection

Continental fuel injection (Fig. 3-3 and 3-4) is of the multi-nozzle, continuous flow type which controls fuel flow to match

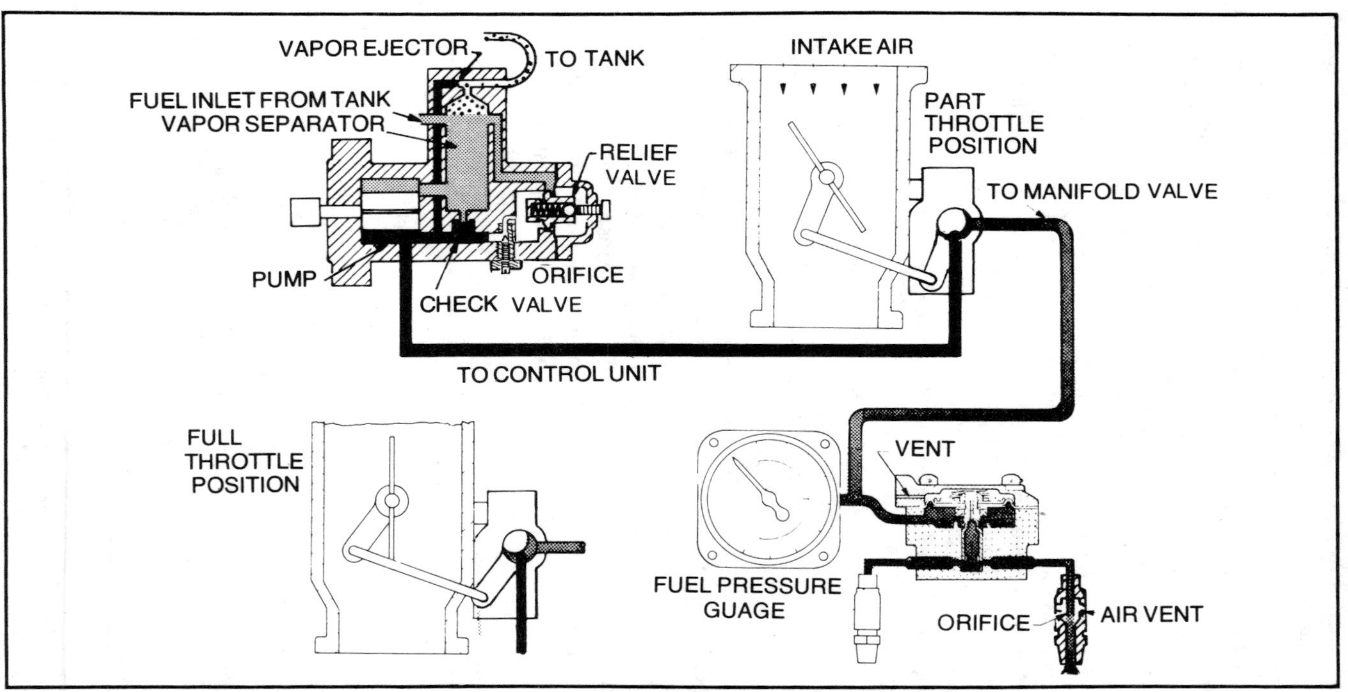

Fig. 3-3. Diagram of the Continental fuel injection system (courtesy of Teledyne Continental Motors).

engine air flow. Any change in air throttle position, engine speed, or a combination of both, causes changes in fuel flow in the correct relation to engine air flow. A manual mixture control and a pressure gauge provide precise leaning at any combination of altitude and power setting. Because fuel flow is directly proportional to metered fuel pressure, settings can be predetermined and fuel consumption can be accurately predicted.

Fuel enters the system at the swirl well of the vapor separator. Here, vapor is separated by a swirling motion so that only liquid fuel is fed to the pump. The vapor is drawn from the top center of the swirl well by a small pressure-jet of fuel and is fed into the vapor return line. This line carries the vapor back to the fuel tank. Ignoring the effect of altitude or ambient air conditions, the use of a positive-displacement engine-driven pump changes pump flow proportional to engine speed. The pump provides greater capacity than is required by the engine and as a result a recirculation path is obtained. By arranging a calibrated adjustable orifice and relief valve in this path, the pump delivery pressure is also maintained proportional to engine speed. These provisions assure proper pump pressure and delivery for all engine operating speeds. A check valve is provided so that boost pressure to the system can bypass the engine-driven pump in starting.

The function of the fuel-air control unit is to control engine air intake and to set the metered fuel pressure for proper fuel-air ratio. The air throttle is mounted at the manifold inlet and its butterfly valve controls the flow of air to the engine as positioned by the throttle control in the aircraft. Main fuel enters the control unit through a strainer and passes to the metering valve. This rotary

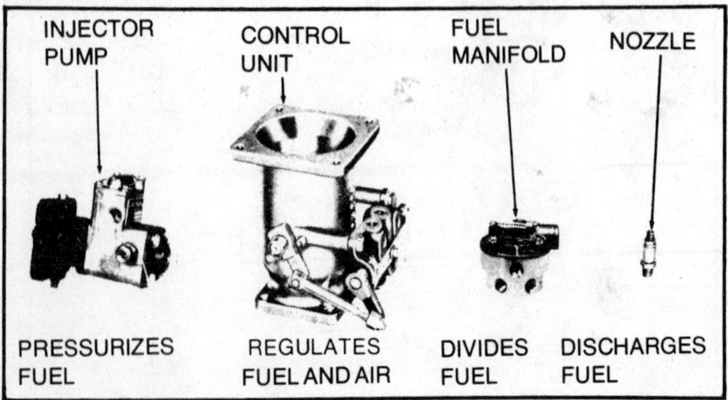

Fig. 3-4. Basic components of the Continental fuel injection system (courtesy of Teledyne Continental Motors).

valve has a camshape edge across the fuel delivery port. The position of the cam at the port controls the fuel passed to the manifold valve and the nozzles. By connecting the metering valve to the air throttle, the fuel flow then is properly proportioned to air flow for the correct fuel-air ratio.

From the fuel injection control valve, fuel is delivered to the fuel manifold valve which provides a central point for dividing fuel to the individual cylinders. In the fuel manifold valve, a diaphram and plunger valve raises or lowers (by fuel pressure) to open or close the individual cylinder fuel supply ports simultaneously. A check valve under the plunger serves to insure that the plunger fully opens the outlet ports before fuel flow starts. As a result, there is no unblanced restriction to flow in the fuel manifold valve.

The fuel discharge nozzle is located in the cylinder head. Its outlet is directed into the intake port. The nozzle body contains a drilled central passage with a counterbore at each end. The lower end is used as a chamber for fuel-air mixing before the spray leaves the nozzle. The upper bore contains a removable orifice for calibrating the nozzles. Near the top, radial holes connect the upper counterbore with the outside of the nozzle body for air admission. These holes enter the counterbore above the orifice and draw outside air through a cylindrical screen fitted over the nozzle body. This keeps dirt and foreign material out of the interior of the nozzle. A press-fitted shield is mounted on the nozzle body and extends over the greater part of the filter screen, leaving an opening near the bottom. This provides both mechanical protection and an air path of abrupt change of direction as an aid to cleanliness. Nozzles are calibrated in several ranges and all nozzles furnished for one engine are of the same range identified by a letter stamped on the hex of the nozzle body.

Mixture control operation is accomplished by a cam-and-doubleport valve which reduces the fuel pressure applied to the metering valve. It permits the operator to lean the mixture for most economical power at any operating point. Provisions are made for a gauge to read metered fuel pressure. This gauge permits simple and accurate adjustment of the mixture for any altitude and power within engine ratings. When the mixture control valve is placed in idle cut-off position (full lean), the fuel supply to the system is simultaneously bypassed back to the inlet side of the fuel pump and flow is cutoff to the metering valve. In this process, the fuel manifold valve closes the lines to each nozzle so that the complete system is primed for easy starting.

Turbocharging

Turbocharging (Fig. 3-6) an engine, a relatively new concept to civil aviation, is a means of overcoming the normal power dropoff of a conventional engine as altitude is increased. A naturally aspirated engine (Fig. 3-7) will produce power in direct proportion to the density of the intake air. This means an engine which delivers 100 horsepower at sea level will deliver about 74 horsepower at 10,000 feet.

Airplanes engines equipped with turbochargers to regain power loss due to altitude are referred to as "normalized" engines. A normalized engine generally utilizes a waste gate valve to bypass the exhaust gas at sea level so that no turbocharging takes place. As the engine starts to lose power at altitude, the waste gate is gradually closed (either manually or by an automatic control) and the turbocharger compresses the inlet air to sea-level pressure. This allows the engine to deliver essentially sea-level horsepower up to an altitude where the waste gate is completely closed and all the exhaust gas passes through the turbine. As the aircraft continues to climb above this altitude (critical altitude), the engine will lose power at the inverse of the pressure ratio reached at critical altitude. At that point, the turbocharger can no longer

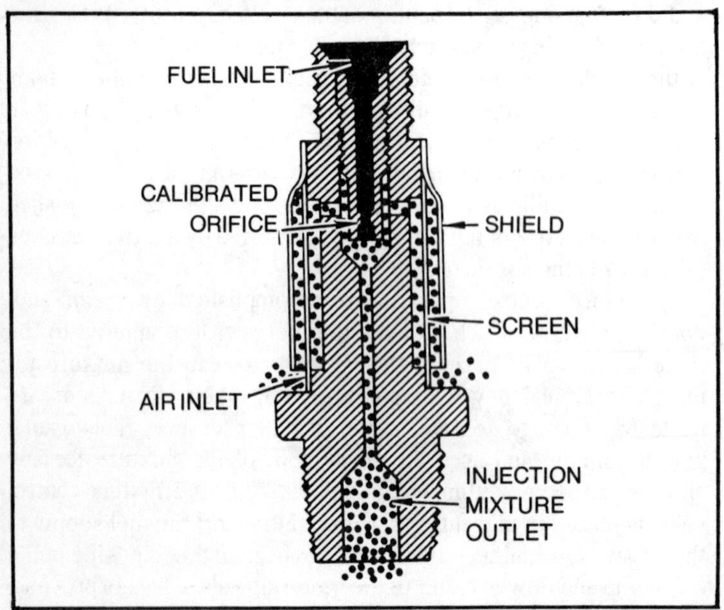

Fig. 3-5. Fuel discharge nozzle (courtesy of FAA).

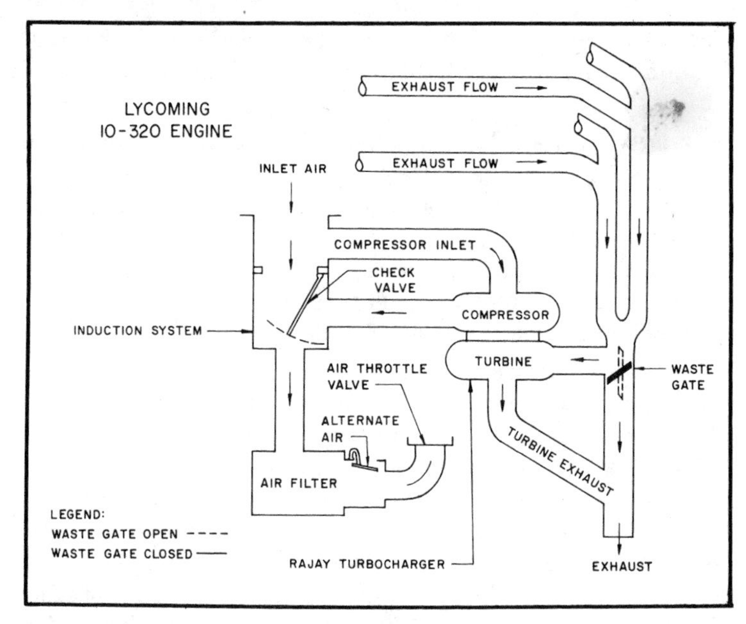

Fig. 3-6. Schematic of the Rajay turbocharging system employed in the Turbo Twin Comanche (courtesy of Rajay Industries).

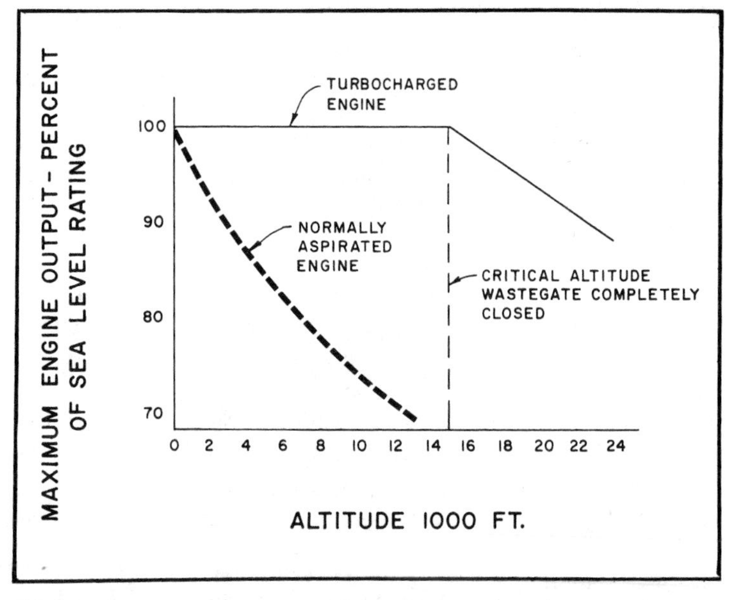

Fig. 3-7. Comparison of a naturally aspirated and turbocharged engine at various altitudes (courtesy of H.P. Books).

deliver air at sea-level pressure. This critical altitude will vary with the engine and the turbocharger used. Principle benefits of turbocharging are:

■ Normal sea level power can be maintained typically to altitudes of 15,000 to 20,000 feet.

■ Additional power can be produced at sea level (for engines so rated).

■ Aircraft ceiling limits are substantially improved; range and TAS are also improved (at high altitudes).

■ Venturi icing is eliminated while turbocharging.

In a properly matched engine-turbocharger installation, engine horsepower is never used to drive the turbocharger; ie., no additional mechanical loads are imposed on the engine. Work required to drive the turbocharger is recovered from the engine exhaust gases; energy that would otherwise be lost is recovered.

A turbocharger, such as the Rajay RJ 0326 used in the Turbo Twin Comanche, is designed to be lubricated by engine lubricant (Fig. 3-8). Oil is supplied to the turbo oil gallery by a line connected to the engine accessory case. A pressure regulator is included in the lubricant supply line to reduce engine gallery oil pressure from 60-80 psi (required for the engine) to 30-50 psi pressure. Between 1 and 2 quarts per minute of lubricant are supplied to the turbocharger. This quantity of oil is a very small percentage of the total engine oil pump capacity. Oil is returned to the engine pump from the turbocharger by way of the oil scavenge pump. A pressure switch is generally incorporated in the turbocharger lubricant supply line to activate a red warning light in the event turbocharger oil pressure is below 27-30 psi.

Engine-Driven Fuel Pump

The function of the engine-driven fuel pump is to deliver a continous supply of fuel at the proper pressure at all times during engine operation. A pump widely used is the positive-displacement, rotary-vane-type pump. The engine-driven pump is usually mounted on the accessory section of the engine. The rotor, with its sliding vanes, is driven by the crankshaft through the accessory gearing. A seal prevents leakage at the point where the drive shaft enters the pump body and a drain carries away any fuel that leaks past the seal. Because the fuel provides enough lubrication for the pump, no special lubrication is necessary.

Because the engine-driven fuel pump normally discharges more fuel than the engine requires, there must be some way of

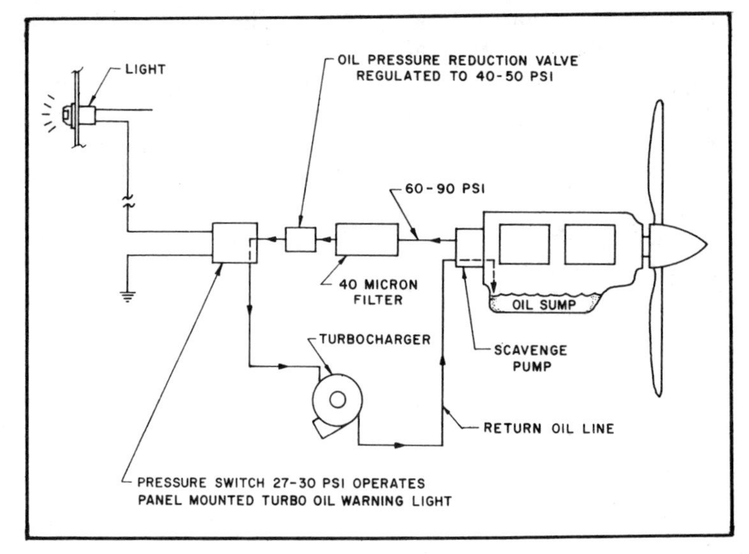

Fig. 3-8. Schematic of the turbocharger lubrication system employed in the Turbo Twin (courtesy of Rajay Industries).

relieving excess fuel to prevent excessive fuel pressures at the fuel inlet of the carburetor. This is accomplished through the use of a spring-loaded relief valve that can be adjusted to deliver fuel at the recommended pressure for a particular carburetor. Adjustment is made by increasing or decreasing the tension of the spring.

In addition to the relief valve, the fuel pump has a bypass valve that permits fuel to flow around the pump rotor whenever the pump is inoperative. This valve consists of a disk that is lightly

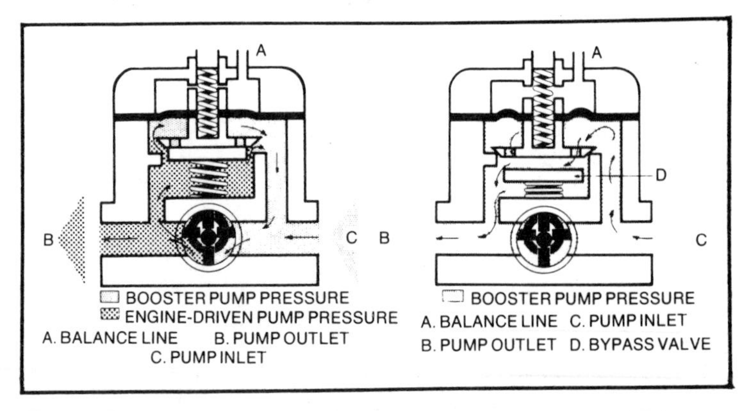

Fig. 3-9. Engine driven fuel pump showing (1) pressure delivery and (2) bypass flow (courtesy of FAA).

spring-loaded against a series of ports in the relief valve head. When fuel is needed for starting the engine, or in the event of engine-driven pump failure, fuel at booster-pump pressure is delivered to the fuel pump inlet. When the pressure is great enough to move the bypass disk from its seat, fuel is allowed to enter the carburetor for priming or metering. When the engine-driven pump is in operation, the pressure built up on the outlet side of the pump, together with the pressure of the bypass spring, holds the disk on its seat and prevents fuel flow through the ports.

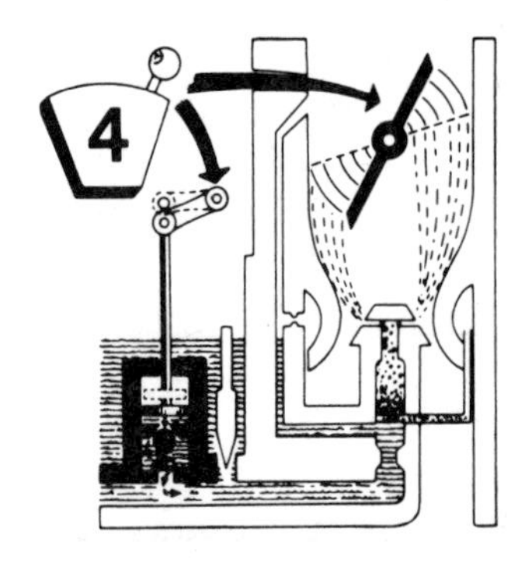

Electrical Systems

Aircraft engines depend upon two basic electrical systems for operation: the magneto-spark plug system for ignition and the alternator-battery system for powering the engine starter motor, auxilliary fuel pump, and electrical accessories. Both of these electrical systems have achieved a high state of perfection.

Early magnetos weighed as much as 20 pounds, utilized large stationary horseshoe magnets to furnish the necessary magnetic field and carried the condenser, coil and contact points on the rotating armature. A magneto today weighs as little as 3½ pounds, rotates the magnet as the armature and locates the condenser, coil and contact points on the stationary element (stator). See Figs. 4-1 and 4-2.

High-Tension Magneto System

A high-tension magneto system can be divided into three distinct circuits. These are the magnetic, the primary electrical and the secondary electrical circuits. The magnetic circuit consists of a permanent multiple rotating magnet, a soft iron core, and pole shoes. The magnet is geared to the aircraft engine and rotates in the gap between two pole shoes to furnish the magnetic lines of force (flux) necessary to produce an electrical voltage. The poles of the magnet are arranged in alternate polarity so that the flux can pass out of the north pole through the coil core and back to the south pole of the magnet.

When the magnet is in position (see Fig. 4-1A), the number of magnetic lines of force through the coil core is maximum because

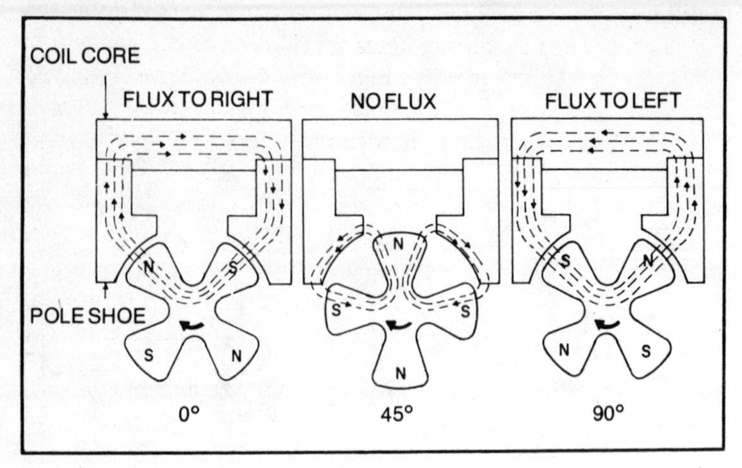

Fig. 4-1. Magneto flux as a function of rotor position (courtesy of FAA).

two magnetically opposite poles are perfectly aligned with the pole shoes. This position of the rotating magnet is called the "full-register" position. As the magnet moves from the full-register position, more lines of flux are short-circuited through the

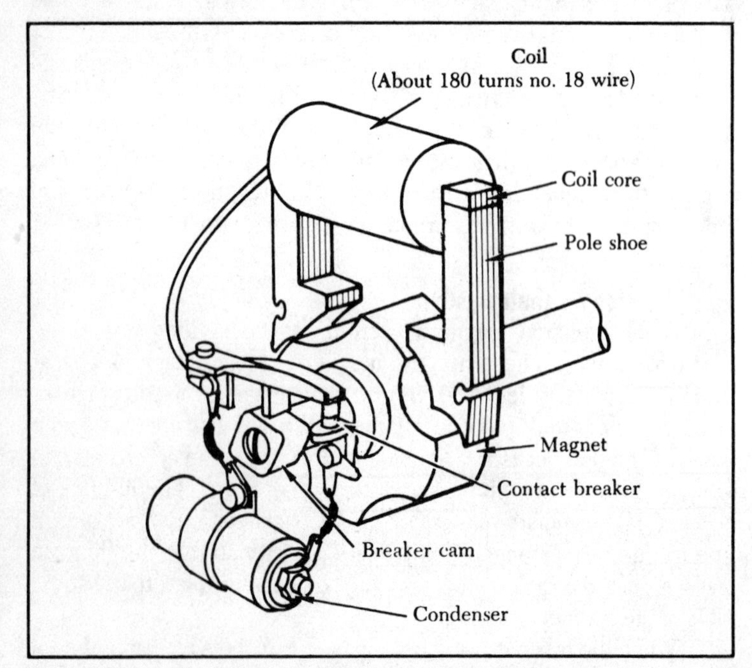

Fig. 4-2. Basic components of the magneto (courtesy of FAA).

pole-shoe ends. Finally, at the neutral position (45° from the full-register position) all flux items are short-circuited, and no flux flows through the coil core (see Fig. 4-1B). As the magnet is moved clockwise, the lines of flux that had been short-circuited through the pole-shoe ends begin to flow through the coil core again. But this time the flux lines flow through the coil core in the opposite direction (see Fig. 4-1C). As a result, for one revolution of the four-pole magnet, there will be four positions of maximum flux, four positions of zero flux and four flux reversals.

When a coil of wire as part of the magneto's primary electrical circuit is wound around the coil core, it is also affected by the varying magnetic field. The primary electrical circuit consists of a set of tungsten breaker contact points, a condenser and an insulated coil. The coil is made up of a few turns of heavy copper wire, one end of which is grounded to the coil core, and the other end connected to the ungrounded side of the breaker points. The primary circuit is complete only when the ungrounded breaker point contacts the grounded breaker point. The third unit in the circuit, the condenser (capacitor), is wired in parallel with the breaker points. The condenser prevents arcing at the points when the circuit is opened and hastens the collapse of the magnetic field about the primary coil. See Fig. 4-30.

The primary breaker closes at approximately full-register position. When the breaker points are closed, the primary electrical circuit is completed and the rotating magnet will induce current flow in the primary circuit. This current flow generates its own magnetic field which is in such a direction that it opposes any change in the magnetic flux of the permanent magnet's circuit. While the induced current is flowing in the primary circuit, it will oppose any decrease in the magnetic flux in the core. As a result, the current flowing in the primary circuit holds the flux in the core at a high value in one direction until the rotating magnet has time to rotate through the neutral position to a point a few degrees beyond neutral. This position is called the *E-gap position* (E stands for "efficiency").

With the magnetic rotor in E-gap position and the primary coil holding the magnetic field of the magnetic circuit in the opposite polarity, a very high rate of flux change can be obtained by opening the primary breaker points and stopping the flow of current in the primary circuit. The sudden flux reversal produces a high rate of flux change in the core. This cuts across the some 13,000 turns of secondary coil (wound over and insultated from the primary coil) and induces a pulse of high-voltage current (approximately 20,000 volts) in the secondary. This in turn fires the spark plug. As the rotor continues to rotate to approximately full-register position, the primary breaker points close again and the cycle is repeated to fire the next spark.

The high voltage induced in the secondary coil is directed to the distributor which consists of two parts. The revolving part is called a *distributor finger* and the stationary part is called a *distributor block*. The rotating part is made of a non-conducting material with an embedded conductor. The stationary part consists of a block also made of non-conducting material that contains terminals and terminal receptacles into which the wiring that connects the distributor to the spark plug is attached.

As the magnet moves into the E-gap position for the No. 1 cylinder and the breaker points just separate, the distributor finger aligns itself with the No. 1 electrode in the distributor block. The secondary voltage induced as the breaker points open enters the distributor finger where it arcs a small air gap to the No. 1 electrode in the distributor block and from that place to the No. 1 spark plug.

Magneto and Distributor Venting

Because magneto and distributor assemblies are subjected to sudden changes in temperature, the problems of condensation and moisture are considered in the design of these units. The high-voltage current that normally arcs across the airgaps of the distributor can flash across a wet insulating surface to ground or the high-voltage current can be misdirected to some spark plug other than the one that should be fired. This condition is called *flashover* and usually results in cylinder misfiring. For this reason, coils, condensers, distributors and distributor rotors are waxed so that moisture on such units will stand in separate beads and not form a complete circuit for flashover.

Flashover can lead to carbon tracking. This appears as a fine pencil-like line on the unit across which flashover occurs. The carbon trail results from the electrical spark burning dirt particles which contain hydrocarbon materials. The water in the hydrocarbon material is evaporated during flashover. This leaves carbon to form a conducting path for current. Even though moisture is no longer present, the spark will continue to follow such a track.

At high altitudes another phenomena occurs. The reduced atmospheric pressure results in longer spark gap distances.

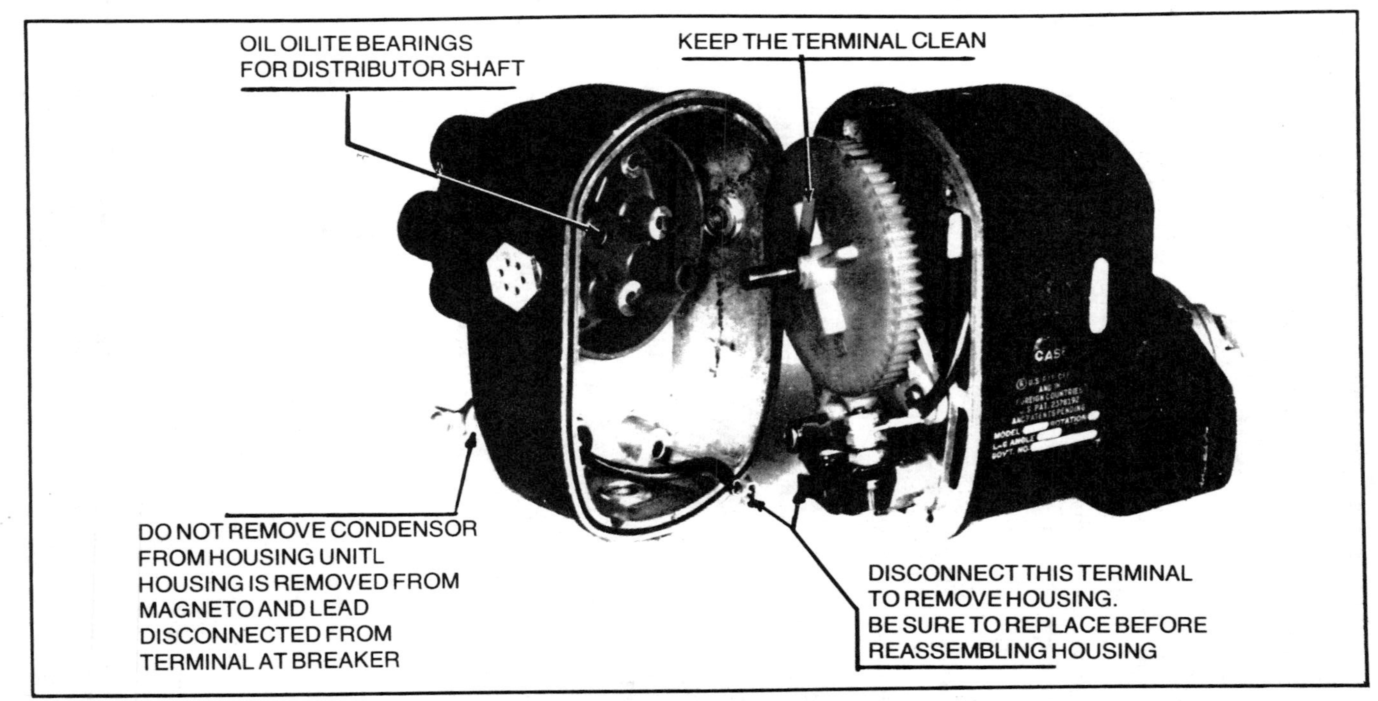

Fig. 4-3. A Slick Electro magneto illustrating disassembly for condensor and point servicing (courtesy Slick Electro Inc.).

Insulating properties of air degrade as pressure is reduced (until a near vacuum is reached). To prevent unwanted arc-over within a magneto, some units are pressurized by turbo bleed for high altitude operation.

Impulse Coupling

Engines having a small number of cylinders are sometimes equipped with what is known as an impulse coupling. This unit will, at the time of spark production, give one of the magnetos attached to the engine a brief acceleration and produce a hot spark for starting. This device consists of small flyweights and spring assemblies located within the housing which attaches the magneto to the accessory shaft.

The magneto is flexibly connected through the impulse coupling by means of the spring so that at low speed the magneto is temporarily held while the accessory shaft is rotated until the piston reaches approximately a top center position. At this point the magneto is released and the spring kicks back to its original position. The result is a quick twist of the rotating magnet. This, being equivalent to high-speed magneto rotation, produces a hot spark.

After the engine is started and the magneto reaches a speed at which it furnishes sufficient current, the flyweights in the coupling fly outward due to centrifugal force and lock the two coupling members together. That makes it a solid unit and returns the magneto to a normal timing position relative to the engine. The presence of an impulse coupling is identified by a sharp clicking noise as the crankshaft is turned at cranking speed past top center on each cylinder.

An alternate technique employs a vibrator and a second set of magneto points instead of a mechanical impulse coupling. When starting the engine, battery current is supplied the vibrator through a second set of magneto points (set at piston top dead center). The vibrator in turn is connected to the magneto coil to produce a hot spark for starting. After engine start, the circuit is disabled and ignition is produced by operating the primary set of magneto points in a conventional fashion. This technique requires that the aircraft battery be capable of supplying an adequate vibrator current when starting the engine. The mechanical impulse system will produce a hot spark for starting without dependence on the aircraft battery.

Aviation Spark Plugs

The aviation spark plug is often thought of as a rather simple thing. Nothing could be further from the truth. This small device must operate under combustion gas temperatures as high as 3,000F, gas pressures as high as 2,000 pounds per square inch and voltages in the order of 20,000 volts. In addition, it must perform reliably in such an environment for hundreds of hours providing tens of millions of ignition sparks and at the same time contain electrical pulses so that there will be no radiation interference to aircraft communications equipment.

An aviation spark plug (Fig. 4-4 and 4-5) consists principally of an outer metal shell, a ceramic insulator (generally aluminum oxide), and a center electrode and ground electrodes, and more often a resistor. Aluminum oxide is most often selected as the insulator because it offers good thermal conductivity, excellent resistance to combustion gas chemical attack, high insulating properties (even at high temperatures) and good heat shock stability. Other materials which will withstand the severe combustion gas corrosion conditions are nickel alloys, platinum alloys and iridium alloys.

Two types of electrodes are used in aviation spark plugs; massive and fine-wire. *Massive center electrodes* are fabricated from a nickel alloy material whereas the *fine-wire electrodes* employ platinum or iridium. Because both of these metals offer low electrical erosion, a very high resistance to chemical attack, and are relatively immune to lead deposition (tetraethyl lead), fine-wire spark plugs which employ these materials for electrodes offer inherently long life and minimum interference with exhaust gas scavenging. Compared to a massive electrode spark plug, a fine-wire spark plug will tend to stay cleaner longer and, if properly cared for, many times will operate until engine overhaul. Unfortunately, the cost of a fine-wire spark plug is approximately three times that of a standard massive electrode plug. To obtain this extended life of fine wire spark plugs, it is very important that the heavy dense lead deposits be removed using a vibrator-type cleaner, followed by light abrasive blasting.

When operating, the ceramic tip of a spark plug must maintain a temperature between two very definite limits (Fig. 4-6). The tip must operate sufficiently hot to prevent lead bromide and carbon foiling and sufficiently cool to prevent preignition. At tip temperatures below 800F, lead and carbon deposits, being conductive in nature, can provide an electrical leakage path from the center electrode to the outer shell. This will short-circuit the spark plug. Above 900F carbon deposits will not form and the temperatures will be sufficiently high to cause the bromide scavenger in aviation gasoline to fully activate on the tetraethyl lead. This will render the disposition of the lead as a gaseous vapor with the combustion exhaust. At tip temperatures between 1,000F and 1,300F, deposits formed on the spark plug are least objectionable and least apt to cause spark plug fouling. At tip temperatures in excess of 1630F, pre-ignition is likely to result.

To obtain optimum tip temperatures in operation, aviation spark plugs are available in a wide variety of *heat ratings* to meet the various engine and operational requirements. The ceramic insulator of a "cold" plug (short core nose) is shaped to maximize heat transfer from the tip area to the outer shell—thus cooling the tip. Conversely, the hot plug features a long core nose and heat path—thus retaining heat in the tip region of the spark plug. The

Fig. 4-4. Massive electrode and fine-wire aircraft spark plugs (courtesy of AC Spark Plug Division, General Motors).

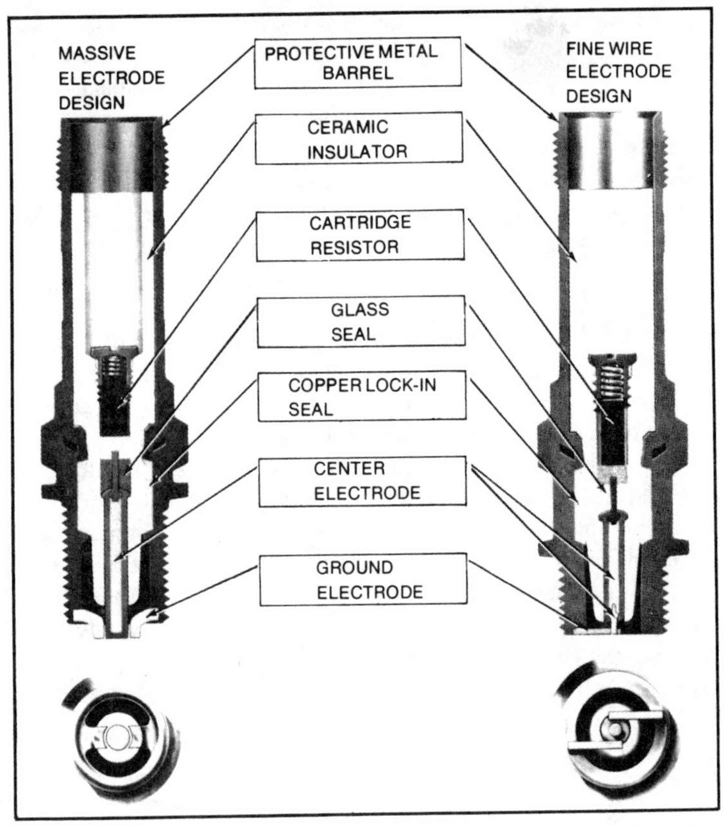

Fig. 4-5. Basic components of an aviation spark plug (courtesy of Champion Spark Plug Co.).

measurement of spark plug tip temperatures over an operating range of idle to takeoff power is performed by the manufacturer using a thermocouple spark plug or similar device. With this knowledge, a precise mating of the engine and spark plug heat ratings is obtainable.

While the amount of tetraethyl lead (TEL) contained in aviation fuel contributes directly to lead fouling of spark plugs, maldistribution of the material plays an even more important part. As strange as it might seem, the amount of TEL distributed to each cylinder in an engine is not necessarily proportional to the amount of fuel delivered to each cylinder. A basic reason for maldistribution of aviation fuel in the aircraft induction system is the difference in the boiling points of the ingredients of aviation fuel. Ethylene dibromide (the lead scavenging agent) has a boiling point of 265F, about 10 percent of aviation fuel hydrocarbons have a boiling point greater than 285F and TEL has a boiling point of about 390F. The difference in boiling points leads to a possibility of fractionalization of aviation gasoline in the induction system. Unvaporized fuel heavy ends, rich in TEL, might go on entirely to one or two cylinders. Where is the more volatile thylene dibromide might be distributed somewhat more uniformly. This means that those cylinders receiving excessive amounts of TEL with insufficient scavenger are likely to form unwated lead deposits on the spark plugs.

It is evident that engine temperature is important in vaporizing all of the components of aviation fuel. Cold weather operation can lead to increased spark plug lead fouling due to a general reduction in induction system temperatures.

A carburetor-type engine operating under conditions of reduced airflow in the induction manifold will tend to form droplets of liquid fuel high in TEL and low in scavenger in the manifold. When ingested by the cylinder, the result is melted beads of lead oxide formed during combustion. If not carried out by the exhaust, these beads tend to drop to the lowest point in the cylinder, the bottom spark plug, and contribute to lead fouling. Some aircraft owners use fine wire-bottom spark plugs to minimize such adverse effects.

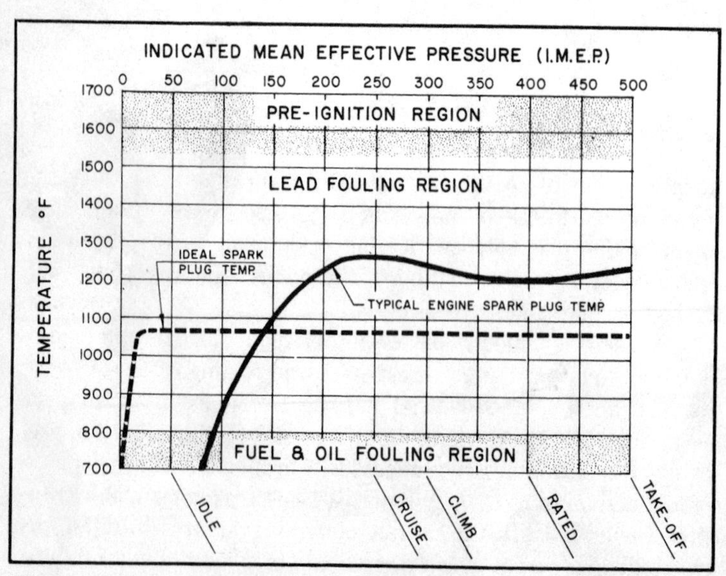

Fig. 4-6. Spark plug tip temperature operating regions.

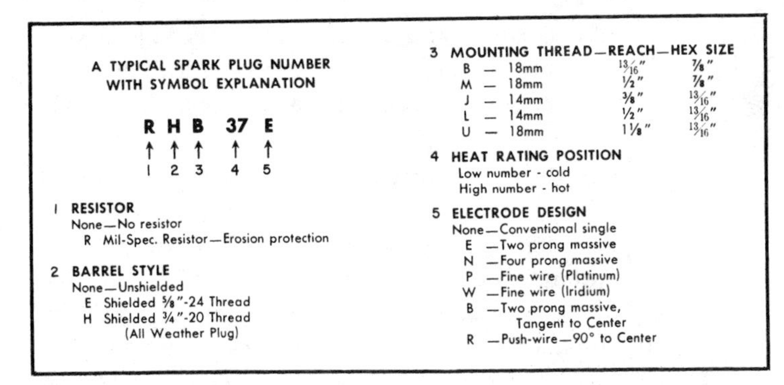

Fig. 4-7. Champion Spark Plug Company designation system (courtesy of Champion Spark Plug Co.).

Fuel injection systems which employ individual spray nozzles to service each cylinder are less prone (but not immune) to maldistribution of TEL. In the typical continuous flow system, the induction fuel-air charge is drawn through the intake valve into the cylinder for approximately one-third of the cycle of operation. For the remaining two-thirds of the cycle, the spray of fuel is directed into the port with no airflow. During periods of long cruise with low power, cold weather or low head temperature operation, liquid fuel, rich in TEL, will form in the manifold and contribute to lead fouling in a manner similar to that of the carburetor engine. Turbocharging improves the distribution of TEL by raising the air temperature within the induction system of the engine.

Spark Plug Inspection and Maintenance

Aircraft owners are permitted by the FAA to perform minor maintenance on their aircraft. Among the things you are permitted to do is to inspect and replace spark plugs. There are two good reasons for performing this maintenance yourself. You will save some money and, examining the story the spark plug can tell is a good way to ascertain the general health of your aircraft engine. The following briefly describes the firing end of a spark plug.

Normal Wear. The firing end shows a light deposit of a brownish-grey material. The general condition of the plug is clean. If the center and ground electrodes show little erosion, the plug can be cleaned, regapped and reinstalled for service.

Gas Fouled. Velvety-black carbon deposits indicate incomplete combustion (usually) due to excessive ground idling or an over-rich fuel mixture. As a means of checking the idle mixture adjustment, note the engine RPM when at idle and move the mixture control idle cutoff. At cutoff, the RPM should hold for a moment and then drop to zero. In the event the RPM increases more than 25 RPM before cutoff, the indication is a rich idle mixture.

Oil Fouled. Black, wet deposits on the lower spark plugs of an opposed engine are normal. However, similar deposits on the upper spark plugs might indicate damaged pistons, worn or broken piston rings, worn valve guides, sticking valves, or a faulty ignition which fails to keep the spark plugs clean. Oil fouling might be encountered during the process of breaking in a new engine.

Lead Fouled. Lead fouling is always present to some degree and can be denoted by a light tan or brown film buildup on the spark plugs' firing end. In severe cases, hard, cinder-like deposits result from poor fuel vaporization, high TEL content in the fuel or an engine that's operating too cold. Using spark plugs of the wrong heat range can also result in lead fouling. Because lead deposits conduct electricity at increased temperatures and pressures, lead fouling promote engine missing during takeoff, METO, and high power cruise operations.

Worn Electrodes. Electrical erosion of spark plugs is, in general, a slow process under normal operating conditions. However, when electrodes have worn to a gap of approximately .030" or half of their original size, the spark plug should be replaced.

To reinstall the spark plugs, first clean the threaded spark plug holes in the cylinder heads with a stiff bottle brush. A thread cleaning tool can be used for brass or stainless steel spark plug busings but not for Heli-Coil insets. Spark plug threads should be coated very sparingly with an approved anti-seize compound, anti-rust, or plain engine oil. The application should begin at least one full thread away from the electrode. Otherwise the compound might migrate to the firing end and foul the spark plug. Spark plugs should first be installed fingertight when torqued to the proper values recommended by the engine manufacturer. Typically, for 14mm spark plugs, a torque value of 19-22 foot pounds is adequate. For 18mm spark plugs, torque values vary from 25-35 foot pounds depending upon the particular engine being serviced.

To install the spark plug lead connections, first clean the terminal connector with an agent such as alcohol, naptha, acetone, or white gas. Install the sleeve without touching the connector or spring. Skin moisture can leave a conducting salt path on the surface which can cause unwanted flashover. Screw on the terminal nut by hand and torque per engine manufacturer recommendation.

As a final installation test, perform a normal magneto check. Do not run the engine on either ignition system alone for more than 30 seconds maximum as the plugs in the shutdown system can become fouled when not operating.

Cold Weather Starting

A phenomenon that can occur in cold weather and cause spark plugs to become completely inoperative is ice bridging of the electrodes. When an engine is shut down in cold weather, moisture can enter the combustion chamber, condense and freeze. In the event ice forms across spark plug electrodes, the device is rendered inoperative by the short circuit. The only cure for this situation is to apply external heat to the engine and melt the unwanted ice.

The Alternator-Regulator System

The heart of the modern lightplane electrical generator system is the alternator and solid-state regulator. While the alternator itself dates back to the days of Edison, the fact that there was no practical way of converting the alternating current output to direct current prohibited its use in small aircraft until the invention of the silicon diode and transistor. The main advantages of an alternator are an intrinsically high output (typically 60 amps) and the ability to produce charging current at engine idle speeds. Light weight, simplicity of construction and precise voltage regulation are additional virtues of the modern aircraft alternator system. The main disadvantage of an alternator is that it will not operate if the aircraft battery is completely discharged. Some aircraft are equipped with auxilliary batteries to supply the alternator in the event of a main battery failure.

There are certain precautions to be practiced in servicing alternator systems. Transistors and diodes used therein are not forgiving of accidental short circuits, reversed polarity or miswiring. *Solid-state devices (transistors and diodes) simply fail completely, instantaneously, and permanently when subjected to accidental electrical overloads*. An aircraft owner or mechanic servicing alternator equipment must acquaint himself with the manufacturers recommended service procedures *before* testing the device.

This section is devoted to a description of Prestolite alternator and regulator equipment as an example of modern lightplane electrical generating and control devices. All comments and service precautions refer to said equipment and are not necessarily applicable to alternator-regulator devices produced by other manufacturers.

The Alternator

As shown in Fig. 4-8, the principle components of a typical Aircraft Alternator are (1) the drive end head, (2) the rotor, (3) the stator, (4) the rectifiers, (5) the slip ring end head and (6) the brush and holder assembly. The drive end head contains a prelubricated bearing, an oil seal (flange mount unit only), collar and shaft seal and a blast tube connection for ventilation. The rotor contains a ventilating fan on the drive end of the shaft (ALX series only) and the slip ring end bearing inner race and spacer on the other end of the shaft. The rotor winding and winding leads are specially treated with a high temperature epoxy cement to provide vibration and temperature resistance characteristics. High-temperature solder is used to secure the winding leads to the slip rings. The stator contains a special lead which is connected to the center of the three-phase windings and is used to activate low-voltage warning systems or relays. The stator is treated with a special epoxy varnish for high-temperature resistance.

Rectifiers used are rated at 150 peak inverse-volts minimum for transient voltage protection. Three positive rectifiers are mounted in the rectifier mounting plate and three negative rectifiers are mounted in the slip ring end head (5″ diameter only). Each pair of rectifiers is connected to a stator lead with high temperature solder. The stator leads are anchored to the rectifier mounting plate with epoxy cement for vibration protection.

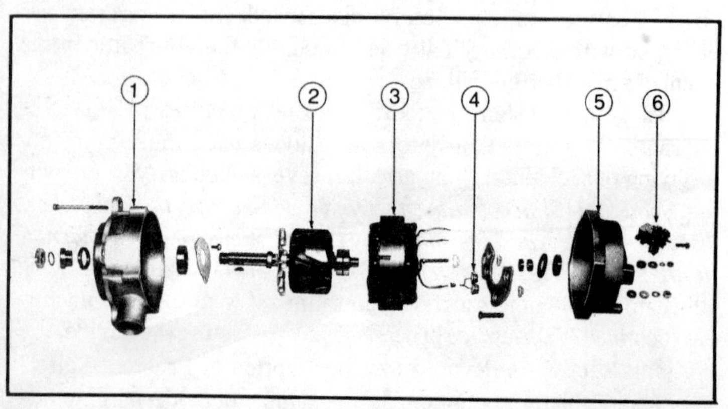

Fig. 4-8. Basic components of the aircraft alternator (courtesy of Prestolite).

The slip-ring end head provides the mounting for the rectifiers mounting plate, output and auxilliary terminal studs, and the brush and holder assembly. The slip ring end head contains a roller bearing and outer race assembly and a grease seal. The brush and holder assembly contains two brushes, two brush springs, and a brush holder and insulators. Each brush is connected to a separate terminal stud and is insulated from ground.

The Voltage Regulator

When the aircraft ignition switch is turned on, battery voltage is applied to the input terminal of the regulator (Fig. 4-9). This in turn is applied to the alternator field coil through power transistor T3. As the engine comes up to speed, the alternator output is rectified and current is fed to the aircraft electrical system (as a result, raising the voltage at the input to the regulator). This voltage is compared to a reference voltage within the regulator (zener diode Z1). When the two voltages compare field current supplied the alternator by T3 is reduced to just the proper amount required to maintain the output alternator voltage at the level determined by comparison to the zener diode reference. The same process maintains system voltage constant during operation. Comparison to the reference adjusts the alternator field current (hence output voltage) according to load demands.

Prestolite Electrical System Service Precautions

■ Disconnect the battery before connecting or disconnecting test instruments (except voltmeter) or before removing or replacing any unit or wiring. Accidental grounding or shorting at the regulator, alternator, ammeter or accessories, will cause severe damage to the units or wiring.

■ The alternator must not be operated on open circuit with the rotor winding energized.

■ Do not, at any time, connect the battery direct to the regulator field terminal.

■ Do not attempt to polarize the alternator. No polarization is required. Any attempt to do so could result in damage to the alternator or regulator.

■ Grounding of the alternator output terminal could damage the alternator or circuit and components.

■ Reversed battery connections can damage the rectifiers, aircraft wiring or other components of the charging system.

Battery polarity should be checked with a voltmeter before connecting the battery. Most aircraft are negative ground.

■ If a booster battery or fast charger is used, its polarity must be connected properly to prevent damage to the electrical system components. Auxiliary power unit voltage settings must match the aircraft electrical system (12 or 24 volts).

■ *Remember:* When an alternator or regulator is installed, the mounting bolts or screws become an electrical connection. They must be clean and properly torqued.

Generators

There are still a lot of light airplanes around that are equipped with generators. Generators aren't as efficient as alternators, but they are reasonably dependable when properly maintained, more forgiving when accidently mishandled and less costly to repair.

It's a good idea to inspect the generator every 100 hours of engine operation. Check the external connections and mounting. Then take off the cover band and inspect the brushes and commutator. If the commutator is dirty, remove the generator from the engine and clean the commutator by holding 00 sandpaper against it with a stick while the armature is turning. (*Never* use emery cloth for this chore).

If the commutator is rough, out-of-round, or has high mica, then it will be necessary to disassemble and have the armature commutator turned down on a lathe.

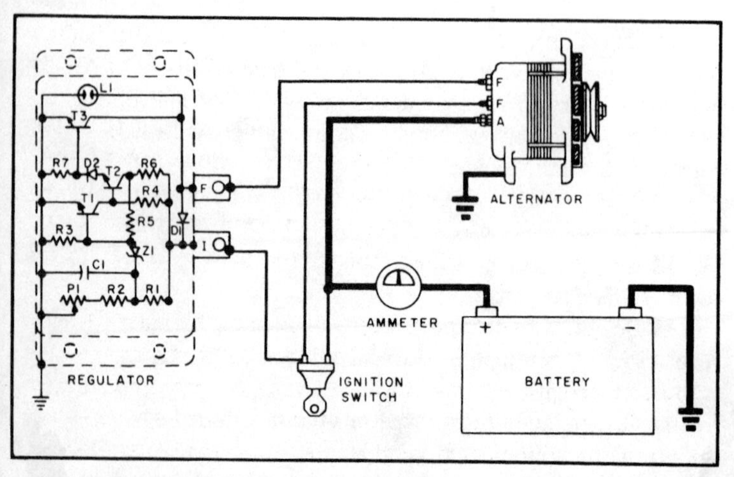

Fig. 4-9. Diagram of the Prestolite aircraft alternator solid-state regulator (courtesy of Prestolite).

When brushes are worn down to less than half their original length they should be replaced. But be sure that the new brushes seat well and that they are in good contact with the commutator. Also be sure that they have proper spring tension.

If the brush springs are blued or burned, it's best to replace them. This is evidence of overheating, which, in turn, means that the springs have lost their temper.

In tracking down generator trouble, the above mentioned should be checked first. A dirty commutator or worn brushes usually cause a low or unsteady output. If the generator is producing nothing at all, then, after all external connections have been checked, and if brushes and commutator appear okay, use a test lamp to find the problem.

Start by raising the grounded brush from the commutator and insulating with a piece of cardboard. Check for grounds from the generator main brush or "A" terminal to the generator frame. If the lamp lights during this test, then the generator is internally grounded. Now, isolate the trouble by insulating all brushes and checking the brush holders, armature, commutator, and field separately. (In case of a grounded field, you'll also probably have to replace the regulator. This condition usually results in an overload that burns the regulator points).

If the generator is not grounded, check the field for an open circuit with the test lamp. The lamp should light when one point is placed on the field terminal or grounded field lead and the other is placed on the brush holder to which the field is connected. If the field is not open, check for a short in the field by connecting a battery and ammeter in series with the field circuit. Refer to the manufacturers data for the appropriate value of field current.

Short circuits in the armature are usually easy to see. The commutator bars will arc each time they pass under the brushes and as a result soon become burned. Shorts in the armature also can be traced with the aid of a *growler*. This piece of equipment can be found in most any auto or airplane generator repair shop.

The Engine Starter

Most lightplanes employ a direct-cranking starting system. The heart of the system is a DC motor. DC motors are the natural choice for a starting system (aircraft or auto) in that they produce an extremely high torque at slow speeds and at stall.

Automatically engaged starting systems employ an electric starter mounted on the engine. A starter solenoid is activated by either a push button or ignition switch on the instrument panel.

When the solenoid is activated, its contacts close and electrical energy energizes the starter motor. Initial rotation of the starter motor engages the starter through a protective overrunning clutch.

Manually engaged starting systems employ a manually operated over-running clutch drive pinion to transmit power from an electric starter motor to a crankshaft starter drive gear. A knob or handle on the instrument panel is connected by a flexible control to a lever on the starter. This lever shifts the starter drive pinion into the engaged position and closes the starter switch contacts when the starter knob or handle is pulled. The starter lever is attached to a return spring which returns the lever and the flexible control to the "off" position. When the engine starts, the overrunning action of the clutch protects the starter drive pinion until the shift lever can be released to disengage the pinion.

Most starting system maintenance practices include replacing the starter brushes and brush spring, cleaning dirty commutators and turning down burned or out-of-round starter commutators. As a rule, starter brushes should be replaced when worn down to approximately one-half their original length. A glazed or dirty starter commutator can be cleaned by holding a strip of double-O sandpaper or a brush seating stone against the commutator as it is turned. The sandpaper or stone should be moved back and forth across the commutator to avoid wearing a groove. Emery paper or carborundum should never be used for this purpose because of possible electrical shorting action. Roughness, out-of-roundness or high-mica conditions are reasons for turning down the commutator on a lathe.

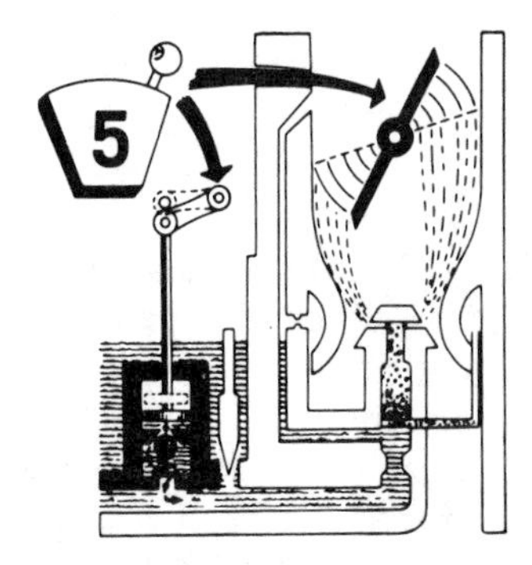

Engine Instruments

The manifold pressure gauge is a primary instrument for setting engine horsepower in aircraft equipped with a constant speed propeller. The measurement of manifold pressure is an indication of the amount of vaporized fuel being fed the engine, hence the output horsepower. RPM is, of course, the other ingredient of horsepower.

Manifold Pressure Gauge

The manifold pressure gauge is calibrated in inches of mercury and measures absolute pressure. When the engine is not running, the manifold pressure gauge records the existing atmospheric pressure. When the engine is running, the reading obtained on the manifold pressure gauge depends on the engine's load. The manifold pressure gauge measures manifold pressure immediately before the cylinder intake ports. The gauge contains an aneroid diaphragm and a linkage for transmitting the motion of the diaphragm to the gauge pointer.

When an engine is not running, the manifold pressure gauge reading should be the same as the local barometric pressure. It can be checked against a barometer or the altimeter in the aircraft can be used. With the aircraft on the ground, the altimeter hands should be set to zero and the instrument panel should be tapped lightly a few times to remove any possible frictional errors. The barometer scale on the altimeter face will indicate local atmospheric pressure when the altimeter hands are at zero. The manifold pressure gauge should agree with the pressure reading.

Engine-Driven Vacuum Systems

The vane-type engine-driven pump (Fig. 5-1) is one source of vacuum for gyros installed in light aircraft. One type of engine-driven pump is mounted on the accessory drive shaft of the engine and is connected to the engine lubrication system to seal, cool, and lubricate the pump. The vane-type pump consists of a case enclosing a rotor mounted eccentric to the case. Vanes mounted on the rotor are free to slide in and out of the shaft as it is rotated.

A typical vacuum system consists of an engine-driven vacuum pump, an air/oil separator, a vacuum regulator, a relief valve, an air filter, and tubing and manifolds neccessary to complete the connections. A suction gauge on the aircraft instrument panel indicates the amount of vacuum in the system. The air/oil separator is a metal box containing a filter screen or a baffle. The oil sticks to the screen or hits the baffle, drops to the bottom of the separator and is drained back into the engine oil pump.

Most vacuum systems contain some type of vacuum regulator which acts as a pressure relief valve in reverse. A spring-loaded valve can be adjusted, typically, for 3.75 to 4.25 in. of mercury. As

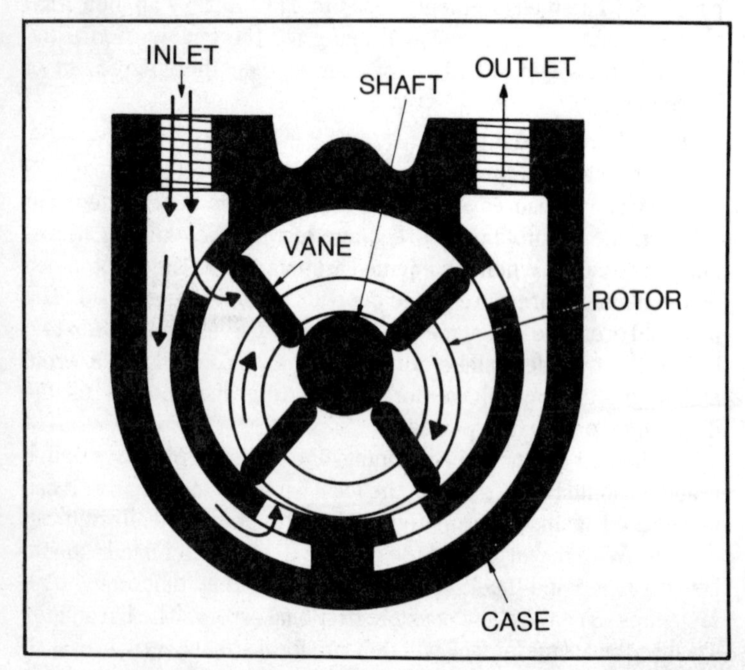

Fig. 5-1. Cutaway view of a vane-type engine driven vacuum pump (courtesy of FAA).

vacuum in the system builds up, it overcomes the spring tension of the regulator and the valve is lifted off its seat. As the valve is lifted, it allows atmospheric pressure to enter and regulates the amount of vacuum developed in the system. A check valve is used in a vacuum system to prevent a reverse flow of air from the pump to the instruments in case of an engine backfire. Air filters prevent foreign matter from entering the vacuum system. Individual filters can be installed for each instrument or a master air filter might be used. This depends upon system design (Fig. 5-2).

Suction Gauge

The suction gauge indicates whether the vacuum system is working properly. The suction gauge case is vented to the atmospheric or to the line of the air filter and contains a pressure-sensitive diaphragm plus a multiplying mechanism which amplifies the movement of the diaphragm and transfers it to the pointer. The reading of a suction gauge indicates the difference between atmospheric pressure and the reduced pressure in the vacuum system.

Tachometers

The tachometer indicator is an instrument for indicating the speed of the crandshaft of a reciprocating engine. There are two types of tachometer systems in wide use today: the mechanical indicating system and the electrical indicating system (Fig. 5-3).

Mechanical indicating systems consist of an indicator connected to the engine by a flexible drive shaft. The indicator contains a flyweight assembly coupled to a gear mechanism that drives a pointer. As the drive shaft rotates, centrifugal force acts on the flyweights and moves them to an angular position. This angular position varies with the RPM of the engine. Movement of the flyweights is transmitted through a gear mechanism to the pointer. The pointer rotates to indicate the RPM of the engine on the tachometer dial. Pointer oscillation can occur with a mechanical indicating system if the flexible drive is permitted to whip. The drive shaft housing should be secured at frequent intervals to prevent whipping.

A typical electric tachometer system utilizes a three-phase AC generator coupled to the aircraft engine and connected electrically to an indicator mounted on the instrument panel. The generator transmits three-phase power to the synchronous motor in the indicator. The frequency of the transmitted power is proportional to the engine speed. Through use of a magnetic drag cup, the instrument furnishes an indication of engine RPM.

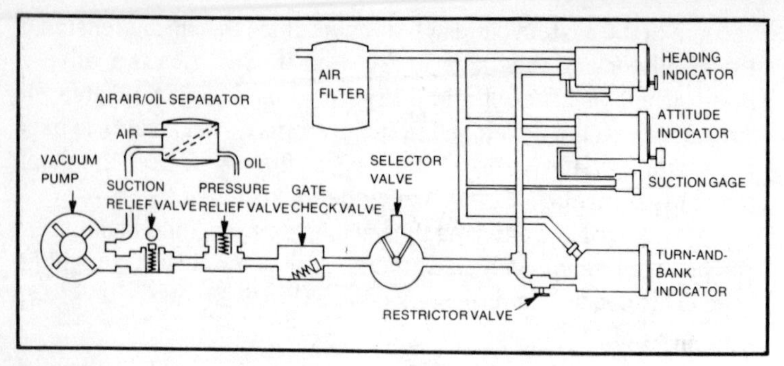

Fig. 5-2. Typical engine driven wet-type vacuum pump system (courtesy of FAA).

Oil Temperature Gauge

Two types of oil temperature gauges are available for use in an engine gauge unit. One type consists of an electrical resistance type oil thermometer that is supplied electrical current by the aircraft DC power system. The other type, a capillary oil thermometer, is a vapor pressure type thermometer consisting of a bulb connected by a capillary tube to a Bourdon tube. A pointer, connected to the Bourdon tube through a multiplying mechanism, indicates on a dial the temperature of the oil (Fig. 5-4).

Cylinder Head Temperature

The cylinder temperature of most air-cooled reciprocating aircraft engines is measured by a heat-sensitive element (thermocouple) attached to some point on one of the cylinders (normally the hottest cylinder). A thermocouple is a circuit or connection of two unlike metals. Such a circuit has two dissimilar metal junctions. If one of the junctions is heated to a higher temperature than the other, an electric current is produced in the circuit. By including a meter in the circuit, this current can be measured. The hotter the high-temperature junction becomes, the greater the

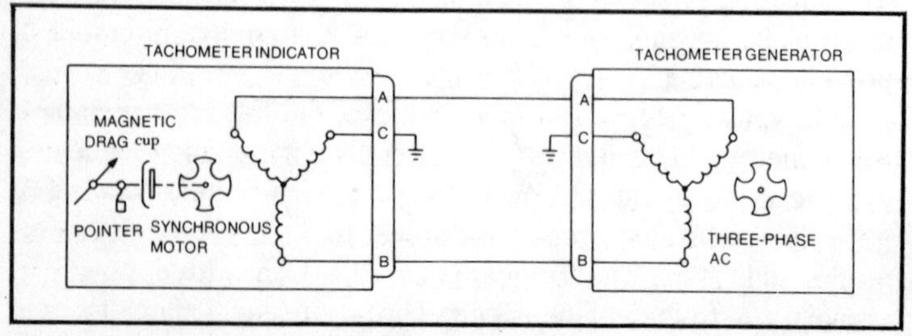

Fig. 5-3. An electrically coupled tachometer system (courtesy of FAA).

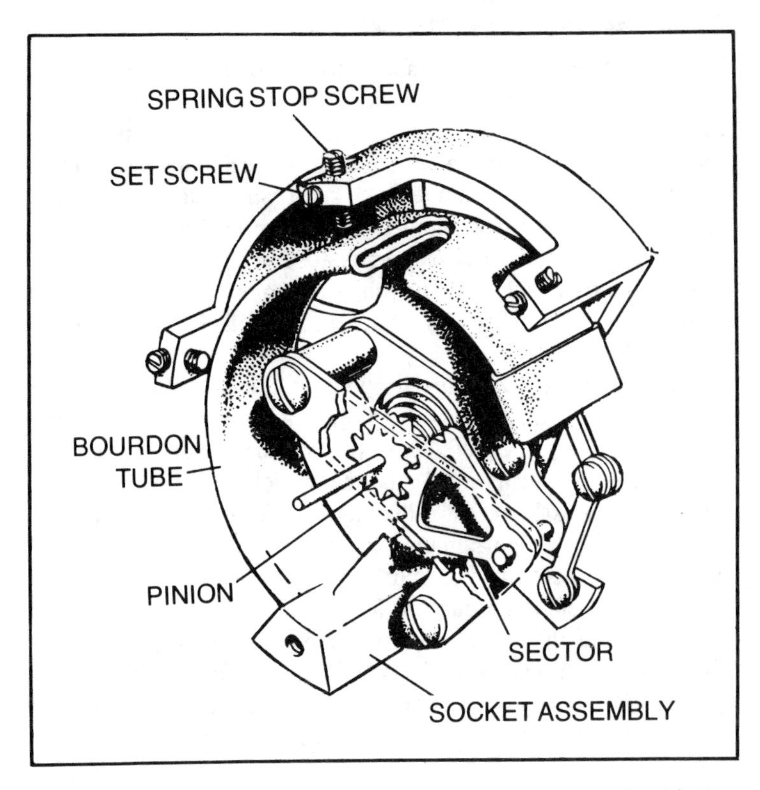

Fig. 5-4. Basic mechanism of a Bourdon tube pressure measurement instrument (courtesy of FAA).

current produced. By calibrating the meter dial in degrees it becomes a thermometer. Thermocouple leads are designed to provide a definite amount of resistance in the thermocouple circuit. As a result, their length or cross-sectional size cannot be altered unless some compensation is made for the change in total resistance.

The hot junction of the thermocouple varies in shape depending on its application. Two common types are the gasket type and the bayonet type. In the gasket type, two rings of dissimilar metals are pressed together to form a spark plug gasket. Each lead that makes a connection back to the meter must be made of the same metal as the part of the thermocouple to which it is connected (typically iron/constantan). The bayonet type thermocouple fits into a hole or well in the cylinder head. Here again, the same metal is used in the lead as in the part of the thermocouple to which it is connected.

Exhaust Gas Temperature (EGT) Indicator

One of the most important factors in flying an aircraft is to maintain the correct fuel-air mixture. Proper mixture to the engine will give maximum range, economical operation and maximum service life. An accurate method for determining the correct fuel-air ratio is a sensitive and fast-responding exhaust gas temperature indicator. Equipment manufactured by Alcor Aviation Inc., San Antonio, Texas is representative of modern aircraft EGT instrumentation and is found on many models of single- and twin-engine aircraft ranging from the Cessna 172 through the Beech Queen Air.

Figure 5-5 shows what happens to the power output, fuel economy (range) and exhaust temperature when fuel-air ratio (lbs. of fuel/lb. of air) is varied. The exhaust temperature curve has a peak at the chemically ideal mixture of fuel and air, 0.067. This mixture gives 100 percent utilization of both the air and fuel. As indicated in Fig. 5-5, maximum exhaust temperature mixture gives maximum fuel economy with minimum loss in power.

It is the optimum mixture for cruise for an unsupercharged engine where the engine is not cooling or detonation limited. Cessna at one time defined "normal lean" mixture as that mixture which gives a 2-mph loss in airspeed when the mixture is leaned from best power mixture. This mixture gives maximum exhaust gas temperature. See Fig. 5-6.

The EGT method of mixture control is accomplished by a fast responding sensing probe located in the aircraft exhaust system and a temperature indicating meter as the pilot readout. Continental and Lycoming recommend that the exhaust probe be located in the exhaust of the leanest cylinder. If the probe is located at the cluster of cylinders then the leanest cylinder will be on the lean side of peak when the cluster setting is at peak or in some cases even 50 F below the peak on the rich side. When a cylinder operates on the lean side of peak EGT at high cruise powers, the resultant oxidizing exhaust can cause excessive deterioration of some valve materials which can lead to engine failure from preignition (See Fig. 5-7). Although the engine manufacturers define the leanest cylinder for some engine models, this does vary—especially with carburetor engines. The only practical way to determine the leanest cylinder is to have an exhaust probe for each cylinder with a selector switch so that the EGT for all cylinders can be scanned at any time to pick the leanest; i.e., the one with maximum EGT. This equipment configuration is referred

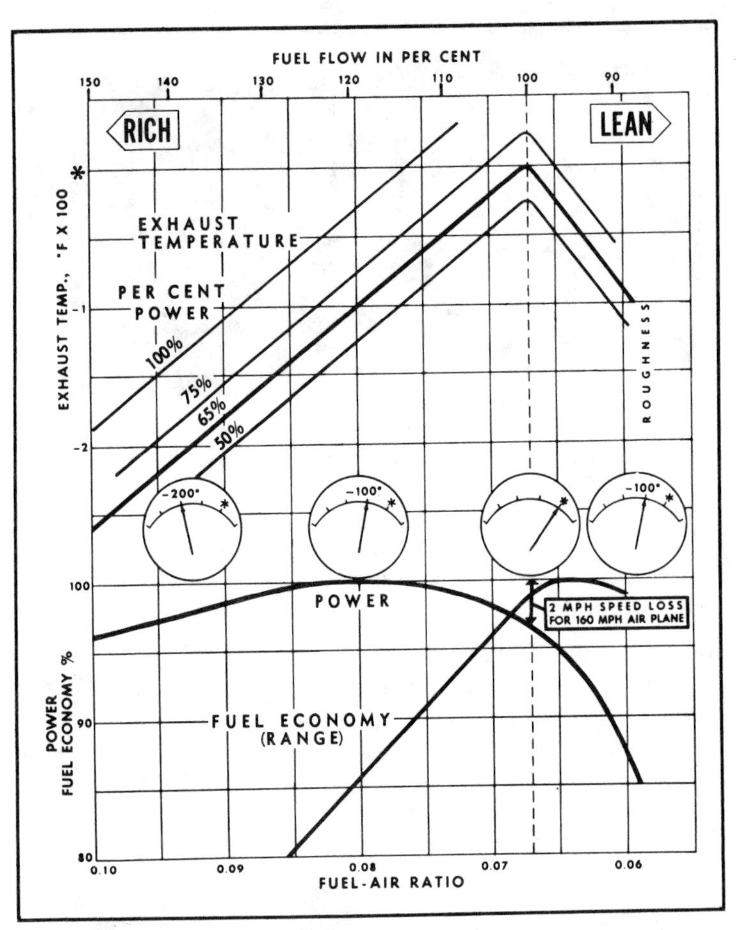

Fig. 5-5. Vital engine parameters in relation to fuel-air mixture (courtesy of Alcor Aviation Inc.).

to as an EGT Engine Analyzer or Combustion Analyzer. See Figs. 5-8 and 5-9. The following briefly summarizes the technique for using an Alcor EGT Engine Analyzer:

Meter Calibration to Reference Temperature

■ Establish cruise flight at an altitude where, at normal cruise RPM, full throttle will produce 65 percent power for the engine(s). (For turbosupercharged engines with automatic waste gate control, use an altitude of about 6500 or 7000 feet above sea level and establish 65 percent power.)

■ Lean mixture to peak EGT. (Use highest reading cylinder).

Fig. 5-6. Exhaust gas temperature gauges for twin- and single-engine aircraft (courtesy of Alcor Aviation Inc.).

■ Set pointer at * or at 4/5 scale if no * on dial. Some meter models have a reference pointer which can be used in lieu of * on dial or 4/5 scale position.

Takeoff and Climb (After Calibration)

■ For unsupercharged engines at full power, the normal EGT is 100 F below reference temperature (*).

■ For highly supercharged engines such as IGSO-540, about 150 F to 200 F below reference temperature is normal.

Fig. 5-7. Piston damage due to preignition. Often resulting from a hot spot within the combustion chamber, preignition can destroy a piston or cylinder head in a matter of minutes. Preignition can be caused by a cracked spark plug insulator, exhaust valve blow-by or scale formation within the cylinder head (courtesy of Alcor Aviation and *The Pilot*, AOPA).

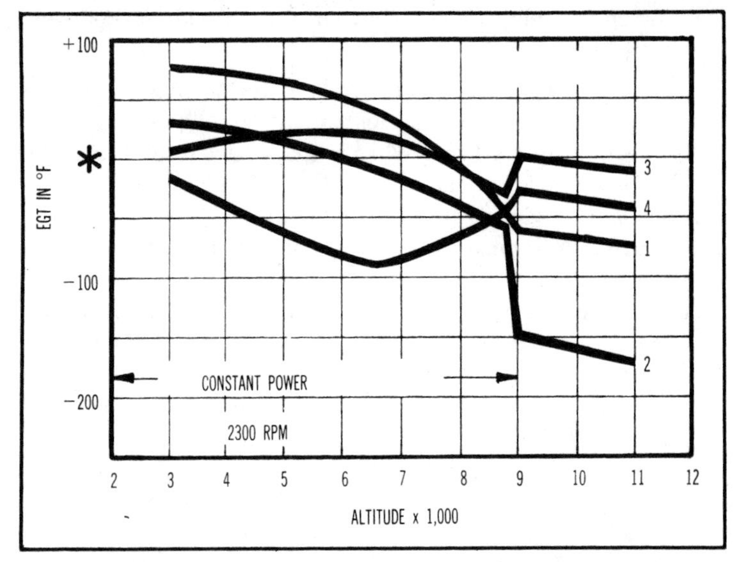

Fig. 5-8. Exhaust gas temperature measurements for individual cylinder in relation to aircraft altitude for a typical four-cylinder engine (courtesy of Alcor Aviation and *The Pilot*, AOPA).

■ Let cylinder head temperature dictate; i.e., enrich mixture and/or increase airspeed to keep engine cylinder head temperature within limits —preferably not over 400 F.

Cruise

■ Select cylinder with highest EGT. Lean mixture slowly enough for pointer to follow. EGT will first increase and then decrease. The peak or point of maximum needle deflection is the reference EGT.

■ Mixture is in accordance with Teledyne Continental recommendations when enriched so that the EGT is at least 50 F below peak.

■ Operation at peak EGT up to 75 percent cruise power for unsupercharged Lycoming engines and up to 65 percent power for supercharged Lycoming engines is approved by Avco Lycoming.

Descent

■ Enrich mixture to drop EGT about 50 F to 100 F before reducing power.

■ While descending, continue to enrich mixture as necessary to keep EGT at peak or on rich side of peak.

■ During prolonged descent, maintain sufficient power to keep EGT on scale (midscale preferably). This will assist in keeping engine temperatures above minimums and prevent over-cooling.

An EGT Engine Analyzer permits detecting engine troubles through a change in exhaust gas temperature as follows:

Spark Plug Failure. When a spark plug stops firing, the EGT of that cylinder will increase 75 F to 100 F. Intermittent firing will result in a lower increase in temperature.

Faulty Magneto And Timing. Complete failure of a magneto will result in the same increase in EGT as failure of a spark plug and this change will be observed for all cylinders. Faulty ignition, such as bad breaker points, will show a change in EGT for the bad magneto relative to the good magneto. A malfunctioning that results in a change in spark timing will result in a pronounced change in EGT. Advanced ignition will decrease the EGT whereas retarded ignition will increase the EGT. Comparing the EGT for single magneto operation will show whether both magnetos are timed the same (ground test only, of course).

Exhaust Valve Failure. Exhaust valve leakage by guttering or warping will result in increasing the EGT. When sufficient

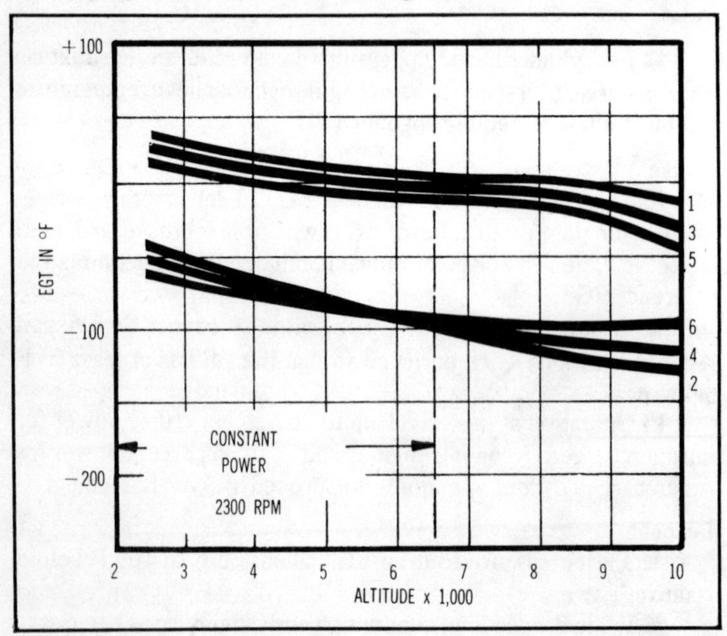

Fig. 5-9. Exhaust gas temperature measurements for individual cylinder in relation to aircraft altitude for a typical six-cylinder engine (courtesy of Alcor Aviation and *The Pilot*, AOPA).

leakage occurs to show a significant increase in EGT, the exhaust valve will deteriorate rapidly with time and, if corrective action is not taken, engine failure could result. This sometimes manifests itself by the valve head breaking off and going through the piston, etc.

Detonation and Preignition. Detonation results in decreasing the EGT because of the increased heat transfer to the cylinder head. Detonation can be distinguished from other EGT-reducing phenomena by the broadness of the peak. When detonation is present the EGT will not increase with mixture leaning. Detonation can lead to preignition. That will normally cause the EGT to increase continuously until engine failure results. Increases in EGT of over 500 F are not uncommon with preignition. The common cause of detonation with today's engines is refueling with low octane fuel in an aircraft requiring high octane. ALCOR received one such report where one aircraft tank requiring 100/130 was filled with 80/87 inadvertently. After establishing cruise conditions, the pilot on leaning the mixture observed a 75 F lower then normal peak EGT with the peak being very broad. In switching to the other tank, to which no 80/87 had been added, normal EGT was obtained. If the pilot had continued to operate under knocking conditions, the severity was such that preignition probably would have occurred (with resultant engine failure).

Fuel Distribution. Poor fuel distribution will result in large changes in EGT from cylinder to cylinder. Fuel injection engines have good fuel distribution as a rule, but a bad injector can cause a cylinder to run so lean as to cause engine failure. Carburetor models of horizontally opposed engines can have such poor fuel distribution that the EGT between cylinders for a given mixture setting can vary more than 200 F. Fuel distribution can be accurately determined by comparing, for the individual cylinders. The mixture setting maximum EGT.

In addition, an EGT indicator is of value in preventing trouble from exhaust gas toxicity. The leakage of exhaust gases into aircraft cabins has caused many catastrophes. When a pilot suspects leakage of exhaust gases into the cabin, the danger of carbon monoxide poisoning can be reduced by leaning the mixture. At peak EGT and leaner, the carbon monoxide content is essentially zero. But at a mixture setting 100 F below peak EGT on the rich side (best power) carbon monoxide constitutes 6 percent of the total exhaust gas.

Carburetor Temperature Indicator

Carburetor ice! Chances are if you fly a carburetor-equipped aircraft you have already encountered the condition—possibly several times. If not, expect to as you build flying time. With all the reliability and simplicity of a carburetor engine, there remains one classical defect: the tendency to ice. As strange as it may seem, there is no foolproof instrumentation that will indicate "carb ice" to a pilot. An instrument that goes a long way toward this goal, however, is the carburetor temperature indicator. One such device is the Richter Aero Equipment Type 8-5 probe and indicator (Fig. 5-10).

To form ice in the throat of a carburetor requires a *freezing temperature and adequate moisture*. The indication of temperature alone does not supply information concerning the presence of sufficient moisture to form ice. Experiments indicate that humidity is the controlling factor in the rate of icing. Therefore, the more humid the air, the more rapid the icing. Dewpoint indications given by FSS weather reports are a fair indicator of moisture in the air.

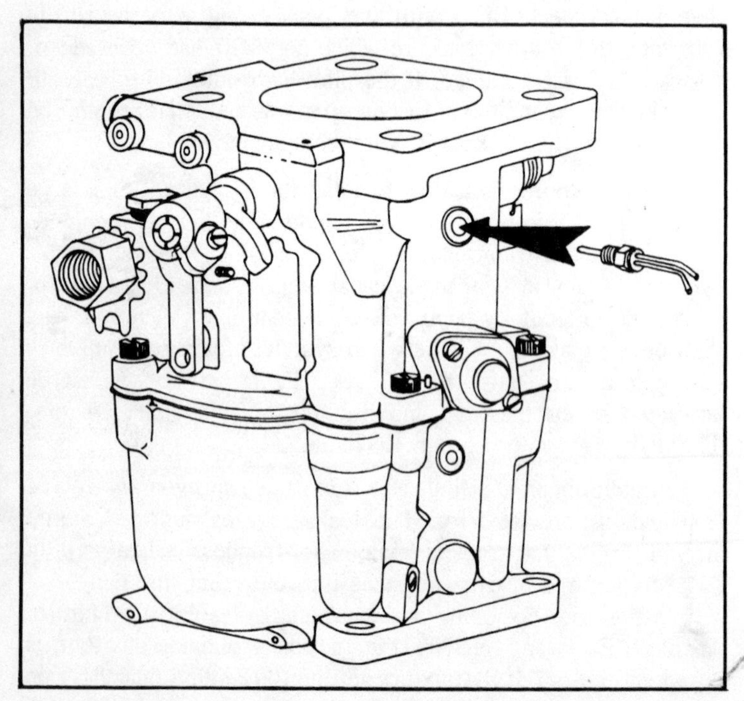

Fig. 5-10. Installation of the Richter Aero Equipment 8-5 probe in the Marvel-Schebler MA4-5 carburetor (courtesy of Richter Aero Equipment).

The closer the dewpoint to the reported temperature, the higher the humidity. On the other hand, it is quite possible to fly ice-free with temperatures 30 degrees to 50 degrees below freezing.

Ice formation in carburetors seems to give its principal trouble at or near the actual freezing point where moisture, condensing on cold metal, begins to build up a deposit. This usually starts adjacent to the throttle valve. It is an expansion-refrigeration effect that manufactures carburetor ice from moist air. The pilot must be alert to keep the carburetor heat level above freezing during conditions of high humidity. If allowed through oversight to drop a degree or two below freezing, the partial use of carburetor heat can bring about exactly the kind of icing trouble a carburetor temperature installation is designed to avoid.

In practice, it is generally sufficient to carry 5 C of indicated heat above freezing under all but the most extraordinary conditions (such as extreme icing). Frequent monitoring of the gauge is required during icing conditions. Induction system icing can occur at several points. The intake screen can become locked with frozen moisture, that is either in the form of sleet or heavy snow. Elbows where the air box angles sharply can be rammed full of incident ice. And most commonly, the throttle valve can accumulate a rim of ice which, if allowed to develop unchecked, will eventually grow to join a deposit which usually forms first on the wall of the carburetor barrel adjacent to the throttle valve. This is the point where the carburetor temperature sensing probe is located. The alternate air supply via the carburetor heater will enable continued operation of the engine even when the intake screen is blocked. If the obstruction at the throttle valve grows large enough to cut off much of the air supply, no alternate source is available and engine stoppage will result.

A collateral benefit derived from the use of a carburetor temperature gauge has come to light as a result of complaints about plug fouling in high compression engines. As explained earlier, a major cause of spark plug lead fouling results from poor vaporization of TEL, ehtylene dibromide and fuel heavy ends. To avoid this, warming the fuel-air mixture in the carburetor aids the volatilization of all the fuel elements.

The intake manifolds of aircraft engines are "tuned" to the specific density of fuel-air mixture which occurs at +5 C. As each cylinder, in turn, sucks a charge from the manifold, shock waves reverberate up and down the manifold. If the manifold is "in tune," each cylinder will pick up an equal charge. If not, one cylinder will run rich and another will run lean. The result is unequal work being performed and unequal distribution of stress and temperatures over the whole engine.

Experiments conducted by Richter Aero using all-cylinder temperature monitoring equipment have shown that, without carburetor heat, cylinder temperature variations of 100 F are common. By using carburetor heat to maintain a temperature of + 5C at the throttle valve, cylinder temperature variations are reduced to 30 F or less. Richter reports obtaining 1742 hours on an 0-470J engine with compression still good, oil consumption normal and Spectra-Chek fine. The use of carburetor heat to optimize carburetor temperature (hence mixture density) was standard operating practice throughout the life of the engine.

Leaning the mixture to compensate for the slight enrichment due to heated carburetor air should result in fuel economy equal to or even better than that experienced when fuel is mixed with very cold air. This applies to cruise power conditions. For maximum power, the densest available, hence coldest, air is required.

The carburetor temperature indicator is fundamentally an electrical resistance thermometer. A heat sensitive element located in the temperature sensing probe is connected electrically as one leg of a balanced bridge. When the resistance of the sensing probe changes, due to a temperature change, there results a change in current through the meter arm of the bridge. This results in an indication of temperature to the pilot. The use of a bridge circuit makes the temperature measurement basically insensitive to aircraft system voltage.

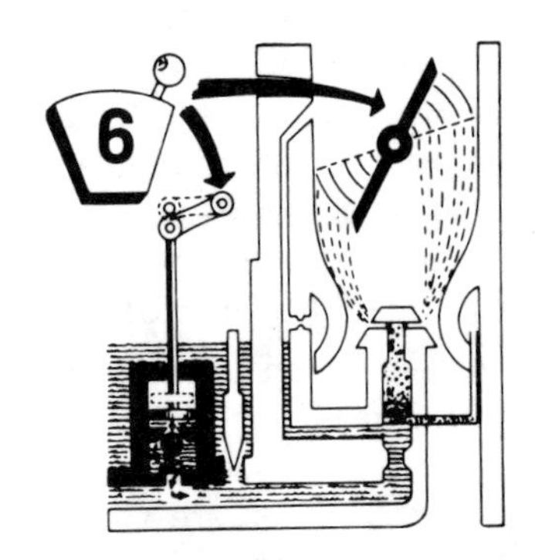

Engine Operation

Engine operation is the subject of many hangar conversations, magazine articles and aircraft owner handbooks. No text on engines could be complete without at least a brief treatment of the matter. Material here represents some of the lesser known, but very important, engine operational considerations and procedures. To begin, let us consider the life blood of an engine—aviation oil.

Aviation Oils

For many years, petroleum base, non-additive oils were used to the virtual exclusion of all others in both horizontally opposed and radial piston engines. In the early 1950s, following a trend of general acceptance in automotive crankcase lubricants, aviation oils were put on the market which incorporated the same or similar metallic detergent additives commonly used in automotive oils. In retrospect, this effort must now be regarded as generally unsuccessful. This is primarily because of the sensitivity of aircraft engines to preignition and the fact that the barium and calcium metals of these additives contributed unfavorably to the preignition problem in these high output, air-cooled engines.

In recent years, both the all-mineral (non-additive) and the metallic detergent-treated aviation oils are being replaced rapidly by a new class of product. These new oils are commonly referred to as the "ashless-dispersant" type. They are distinguished from the previously marketed metallic "detergent" aviation oils. These oils provide a unique combination of resistance to lacquer, varnish and sludge deposits. At the same time they minimize the oil's contribution to combustion chamber deposits. Inherent in these oils are a certain degree of multi-grade properties which expand an individual oil's applicability over a wider range of ambient temperatures. The military services are purchasing such oils to the exclusion of all-mineral oils for all piston engine aircraft. Airlines have made similar changes to gain engine overhaul life and increased inflight reliability. Gone also is their troublesome problem of rocker-box coking. Likewise we see a rapidly developing trend in the private aviation oils indicating the private plane customer's preference for the improved product.

These aviation oils are described by military specifications: MIL-L-6082 for the all-mineral oil and MIL-L-22851 for the ashless-dispersant type. The important oil properties controlled by specification are viscosity, sulfur content, pour point, flash point, corrosion tendencies and carbon residue.

Automotive crankcase oils should not be used in aviation engines and aviation oils should not be used in automotive engines. The ashless dispersant type aviation oils are specifically designed for air-cooled aviation engines. The all-mineral aviation oils do not contain the antiwear and detergent dispersant additives required by modern automobile engines. Conversely, the presence of these additive materials in automotive oils makes them completely unsuitable for aircraft engines. See Fig. 6-1.

From the engine manufacturers standpoint, the subject of lubricants deserves and gets priority attention. Avco Lycoming has the following to say concerning mineral, ashless-dispersant and synthetic aviation type oils:

Mineral Oil. "We prefer this type of oil be used in new, remanufactured or overhauled engines for the first 50 to 100 hours. The reason being that the rings seem to seat quicker. This is particularly true with nitrided cylinders. However, one of its drawbacks is that its viscosity is rather drastically affected by temperature changes (thickens with drop in temperature and thins with increase in temperature). One commonly known disadvantage of this characteristic is the reluctance of the engine to turn over during cold starts. A lesser known cold weather problem is as follows: During a cold start, as the pistons reach the top of their stroke, the rings are compressed due to the choke in our nitrided barrels. However, with congealed oil in the ring grooves, it is difficult for the rings to compress which causes wear at an accelerated pace or even may cause ring breakage.

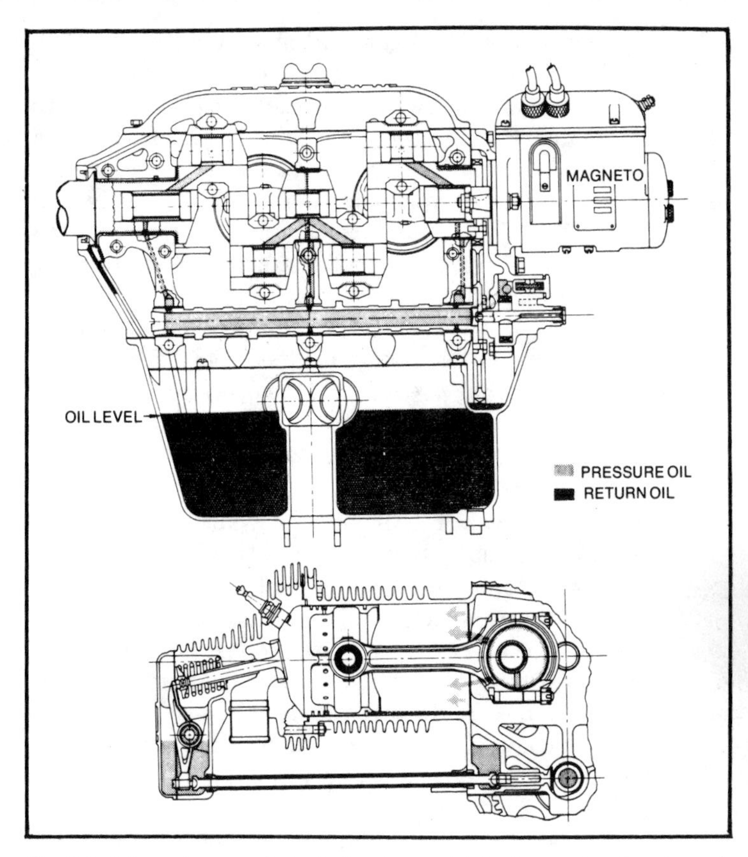

Fig. 6-1. Schematic view of a typical wet-sump lubrication system (courtesy of FAA).

Ashless Dispersant Oils. "This is a mineral type oil with additives or it may be compared to the detergent oil with one important change. This being the removal of the harsh cleansing detergent that would break loose accumulated carbon in the engine which could result in clogged oil screens. This harsh detergent additive was replaced by one that picks up carbon in minute particles and holds them suspended in the oil, preventing the formation of carbon or sludge. Thus, the ashless dispersant oil will keep a clean engine clean, but will not clean a dirty engine as does a detergent type oil. Detergent oils are not FAA approved for aircraft engines.

"Another fine feature of this oil is that it is of the multi-viscosity type. This means that temperature changes do not readily affect the oil viscosity. In simple terms, this gives a starting viscosity in the area of SAE 10 mineral oil while retaining SAE 40 characteristics at operating temperatures. Therefore, the grade 80 ashless dispersant oils can be used the year round in most areas of the United States. For continuous operation in sub-zero temperatures, a grade 65 is available. Avco Lycoming Service Instruction No. 1014C states that 40 weight multi-viscosity oils may be used from 0F to +90F. This is a decided advantage with our current aircraft capability of traveling great distances and encountering wide changes in temperatures.

"Most of the major oil companies have marketed an ashless dispersant oil. Another fine feature of these oils is, and while we don't recommend mixing of oils, one brand may be added to the other if your brand is unattainable or, in a pinch, you could add a quart or two of mineral base oil without harmful effects.

Synthetic Oil. "Currently, we know of only one synthetic oil available—Anderol 471. This is purely a synthetic type oil and possesses excellent lubricating qualities. This lubricant has been recently changed to an ashless type lubricant.

"If one is going to be operating continuously in sub-zero temperatures, this oil is in a class by itself. It has the decided advantage over all other oils for cold starts, as the pour point of this oil is in the area of -40F. Because nothing is all plus, it does cost about $2.65 per quart. In an emergency any other oil is compatible, but the addition of other oils detracts from the purpose of this lubricant.

"Should one decide to use this synthetic oil, please consult Avco Lycoming Service No. L147."

Continental does not currently endorse Anderol due to undesirable effects on certain engine seals. On the subject of a pilot or mechanic adding "super-lubricant" ingredients to aviation oils, Lycoming states the following:

"Avco Lycoming does not approve any additives to the oil used in its aircraft engines. The modern FAA approved aviation lubricants do not require additional additives. Visitors to our factory have observed and commented on the fact that we coat certain moving surfaces of the engine with an additive oil during assembly. Production engineering recommended this procedure only for the initial run of the engine on test and not for continued field use by the owner. The heavy viscosity provides a protective coat on the bare metal for its initial run. However, this same characteristic tends to become a problem with continued use in the field by clogging oil control rings and causing sticking valves."

Aviation oils are classified by weight in a manner similar, but not identical, to automotive oils. The society of Automotive Engineers (SAE) has established the following equivalent oil groupings:

Commercial Aviation No.	Commercial (Auto) SAE No.
65	30
80	40
100	50

Cold Weather Engine Operation

Engine starting during extreme cold weather is generally more difficult than during temperate conditions. Lubricating oil tends to become stiff. This increases the cranking energy required. Prior to operation and/or storage in cold weather, make certain the engine oil is in accordance with manufacturers recommendations.

A fully charged battery can produce an output equivalent to only one-half of its capacity and aviation fuel does not vaporize readily at low temperatures. False starting (failure to continue running after starting) can result in the formation of moisture on the sparkplugs and short-circuit the ignition system. Moisture formed in this manner will necessitate thawing of the engine by removing the plugs or by supplying heat to the engine. For purposes of our discussion on cold weather operation, we'll limit temperatures down to -25F. For tips on engine operation at temperatures colder then -25F, we recomment that you contact your local Alaskan bush pilot.

The use of preheating is recommended when an engine has been cold soaked at temperatures of 10F and below in excess of two hours. Preheating is best accomplished by a high volume hot-air heater. Small electric heaters which warm only the top of an engine can result in superficial preheating—a condition to be avoided. This condition typically results from applying heat to the upper portion of an engine for a few minutes after which the engine is started and normal operation is commenced. Normal cylinder head and oil temperature indications can be achieved. However, oil in the engine sump and oil lines might still be cold enough to congeal. The application of high power, which requires a corresponding free-flow of engine lubricant, might be restricted by congealed oil in the more remote parts of the engine system and give rise to the possibility of engine damage.

Proper procedures for preheating an engine require thorough application of preheat to all parts of the engine. Hot air should be applied directly to the oil sump and external oil lines as well as the cylinders, air intake and oil cooler. Before an engine start is attempted, turn the engine by hand until it rotates freely (typically 8-10 revolutions). This assists in breaking up congealed oil about the various moving parts. When using a single hose preheater, initially apply heat in the lower regions of the engine. Periodically feel the top of the engine and, when warmth is noted, transfer the application of preheat to the upper portion of the engine for approximately five minutes.

Perform a normal engine start and monitor the oil pressure gauge. Because fluid might be congealed in the oil pressure gauge line, as much as 60 seconds could elapse before pressure is indicated. If oil pressure is not indicated in one minute, shut the engine down and determine the cause. Run the engine at approximately 1000 RPM until the minimum allowable oil temperature is indicated. Prior to takeoff, for aircraft with a controllable pitch propeller, cycle the propeller through its operating range three or four times to circulate warm oil through the propeller pitch control mechanism.

The use of commercial "quick-start bombs" is not recommended as an aid to cold weather engine starting. These devices are mainly ether which tends to burn very rapidly. The resultant explosion might be so sharp as to break a ring or cause other engine damage if used in excess.

For a simple preheater, where electricity is available, place a 100-watt lightbulb or so in the lower part of the engine nacelle. Wrap the entire engine nacelle in a blanket and close off the inlet air openings to keep the heat contained within the nacelle. Be certain, however, to contain such lightbulbs in a protective wire cage. Murphy's Law, "if it can go wrong—it will," seems to prevail more often than not. Surely someone will trip over the extension cord; so protect and fuse any simple engine preheaters using lightbulbs.

During cruise, oil temperature should be checked periodically. Should it be lower than normal (below 70C), it might be necessary to block off a portion of the aircraft oil cooler. Normally, oil temperatures at cruise are high enough to evaporate moisture that collects in the oil system. Excessively low oil temperatures permit moisture to collect in the crankcase and rocker box covers. This can result in rust.

Fig. 6-2. Power for many models of Beech, Cessna and Piper aircraft, the very popular Lycoming 10-360 produces 200hp at 2700RPM (courtesy of Avco Lycoming).

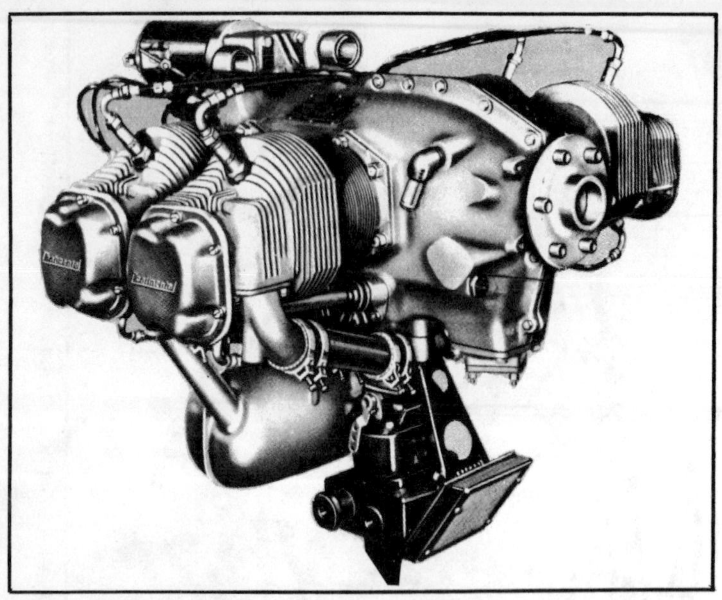

Fig. 6-3. The Continental 0-200 is one of the most popular engine types made. Used in the Cessna 150, the 0-200 produces 100hp at 2750 RPM and weighs 190 pounds (courtesy of Teledyne Continental Motors).

Occasional cycling of constant speed propellers, at approximately 30-minute intervals, during cruise is helpful in preventing congealing of oil in the propeller pitch control mechanism.

Fig. 6-4. Big brother to the 0-200, the Continental 0-300 develops 145hp at 2700 RPM and shares many common parts with the 0-200 (courtesy of Teledyne Continental Motors).

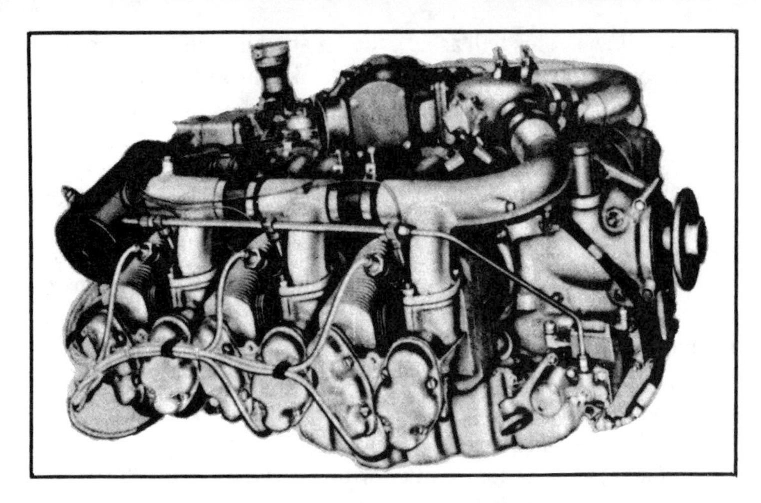

Fig. 6-5. At an engine weight of only 300 pounds, the direct drive turbocharged Continental TS10-360-C produces 225hp at 2800 RPM—power for the Cessna Super Skymaster (courtesy of Teledyne Continental Motors).

Power-off letdowns in extreme cold conditions are to be avoided. When possible, descents should be planned far enough away from the destination that a power letdown can be performed. Should it be necessary to reduce power, for aircraft equipped with an exhaust temperature gauge, the temperature can be peaked

Fig. 6-6. Muscle for the Cessna 180, 182, 188, the Continental 0-470R develops 230hp at 2600RPM. Dry weight is 438 pounds (courtesy of Teledyne Continental Motors).

during letdown. This will assure the greatest possible engine heat for the power setting selected. See Figs. 6-2 through 6-7.

Hot Weather Engine Operation

At ambient temperatures in excess of 90 F, an engine will be particularly difficult to start after having been shut down for approximately 30 minutes to one hour. Although hotter parts such as the cylinders and oil will begin to cool, heat radiated from these parts will be transferred to less temperature tolerant portions of the engine, principally the fuel system. This causes vaporization in the fuel pump and associated fuel lines. During subsequent starting attempts, the fuel pump will be pumping a combination of fuel and fuel vapor. Similarly for fuel injection engines, the injection nozzle line will be filled with varying amounts of fuel and vapor. Until the entire fuel system becomes filled with liquid fuel, difficult starting conditions and/or engine operation will be experienced.

An interesting variable affecting fuel vaporization is the state of the fuel itself. Fresh, high-octane fuel contains a concentration of volatile ingredients. The higher the concentration, the more readily the fuel will vaporize and the more severe the starting problem will become. Time, heat or altitude tends to "age" aviation gasoline; volatile ingredients tend to dissipate. The reduction in volatility could, if a low enough level is reached, result in poor starting due to poor vaporization of the fuel.

To enhance restarting a hot engine, the let down to a destination airport should be conducted at as low a power setting as practical. This will permit the engine to cool. Cowl flaps should be opened fully while taxiing and ground operations are minimized.

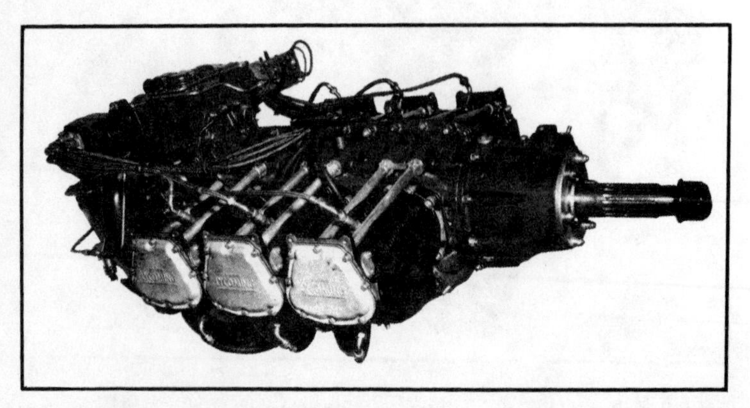

Fig. 6-7. Producing 340hp at 3400RPM, the geared Lycoming GSO-480 powers a wide variety of single- and twin-engine aircraft (courtesy of Avco Lycoming).

Engine Life

Lycoming has studied the frequency of flight and effects on the engine and reports the following:

"We have firm evidence that engines not flown frequently will not achieve the normal expected overhaul life. Engines flown only occasionally deteriorate much more rapidly than those which fly consistently. In view of this, Lycoming accompanies its listed overhaul life in Service Instruction No. 1009 for all models with the statement that the engines must be flown at least 20-30 hours per month and the total time between overhauls must not exceed six years.

Pilots have asked—What really happens to an engine when it's flown only one or two times per month? An aircraft engine flown this infrequently tends to accumulate rust. Some operators are running the engines on the ground in an attempt to prevent rust between infrequent flights. This might harm rather than help the engine if the oil temperature is not brought up to approximately 165 F, because water from combustion will accumulate in the engine oil. The one best way to get oil temperature to 165F is fly the aircraft. During flight, the oil gets hot enough to vaporize the water and eliminate it from the oil.

If the engine is merely ground run, the water accumulated in the oil will gradually contribute to acid—which is undesirable. Prolonged ground running in an attempt to bring oil temperature up is not recommended. Inadequate cooling can result in hot spots in the cylinder, or baked and deteriorated ignition harness, and brittle oil seals causing oil leaks. If the engine can't be flown, then merely pull it through by hand or briefly turn the engine with the starter to coat the critical parts with oil. If the engine is flown infrequently, the oil should be changed at least every 25 hours to eliminate the water and acids."

Engine Storage

One of the most detrimental effects on engine life is long periods of idleness without proper preservative for storage. Moisture condensation, rust, corrosion, sand, and dust—elements that cause pitting of cylinder walls, contamination of oil, and unwanted abrasion—are at work all the time but particularly when an engine is idle. If it is known that an engine will be temporarily out of service for a period of time up to 60 days, the following minimum preservation program is recommended.

■ Remove top and bottom spark plugs and, using a portable pressure sprayer, atomize spray an approved preservation oil through the supper spark plug hole of each cylinder with the piston in the down position (typical preservative oils are Socony Avrex 901, Esso Rust Ban 626, etc.).

■ Rotate the crankshaft as each pair of cylinders is sprayed.

■ Stop the crankshaft so that no piston is in the top position.

■ Re-spray each cylinder without rotating the crank.

■ Re-install the spark plugs (or cylinder dehydrator plugs).

■ Seal all engine openings exposed to the atmosphere using suitable plugs or non-hydroscopic tape.

To remove the engine from storage, remove the seals and tape, hand turn the propeller several revolutions to clear excessive rust preventive oil, and conduct normal startup procedure.

In the event of extended storage, the preceding program of preservation should be repeated at least every 60 days. Depending upon local environmental conditions, a far more comprehensive storage program might be indicated. Tropical and seashore locations require additional protection for the deleterious effects of fungus and salt atmosphere.

For related information on the subject I strongly recommend reading Advisory Circular 43-4, Corrosion Control for Aircraft (5/15/73). This publication summarizes current available data regarding identification and treatment of corrosive attack on aircraft structure and engine materials. It is available free of charge simply by sending a request to the Department of Transportation, Distribution Unit, TAD 484.3, Washington, D.C. 20590.

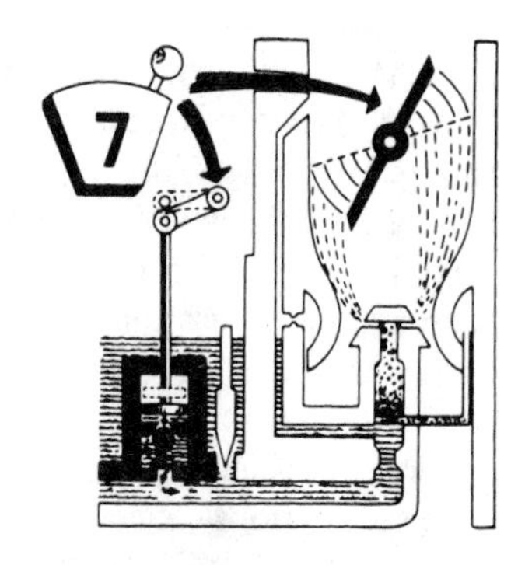

Engine Maintenance

The term *maintenance* applied to an airplane, and consequently an engine, can be subdivided into three categories: inspection, preventive maintenance, repairs and alterations. Federal Aviation Regulation 91.163 states, "The owner or operator of an aircraft is primarily responsible for maintaining that aircraft in an airworthy condition . . . each owner or operator of an aircraft shall have the aircraft inspected as prescribed, etc . . . In addition, he shall insure that maintenance personnel make appropriate entries in the aircraft and maintenance records indicating the aircraft has been released to service." It becomes very clear that the aircraft *owner or operator* is the single individual responsible to have the aircraft inspected, preventive maintenance performed, and repairs or alterations made as necessary to maintain the aircraft in an airworthy condition.

An airplane must have an annual inspection every 12 calendar months. This inspection may be performed by a certified airframe and power plant mechanic holding an FAA Inspection Authorization (IA), an FAA certified repair station or the manufacturer of the aircraft if he meets the requirements of the regulations. An aircraft used to carry passengers for hire, or for flight instruction for hire, must be inspected within each 100 hours of time in service. The annual inspection is acceptable as a 100-hour inspection, but the reverse is not true. Following an inspection, it is the responsibility of the IA mechanic to certify that the aircraft or engine is again airworthy. This is accomplished by entering in the maintenance records a description of the type of inspection, the date of the inspection and the signature and certificate number of the person approving the aircraft for a return to service.

A factor in reducing flying costs is that certain maintenance is not classified as major and may be performed by the pilot himself. Federal Aviation Regulations state, "the holder of a pilot certificate issued under Part 61 may perform preventative maintenance on any aircraft owned or operated by him that is not used in air carrier service." Preventive maintenance means simple or minor preservation operations and the replacement of small standard parts not involving complex assembly operations. Examples of preventative maintenance are replacing and/or cleaning sparkplugs, changing oil, cleaning fuel and oil strainers, replacing hose connections (except for hydraulic lines), replacing prefabricated fuel lines, and the like.

In the event a major repair or alteration is accomplished, it must be inspected and returned to service by an A&P mechanic with an FAA Inspection Authorization, or a properly certificated repair station, manufacturer, air carrier or commercial operator. Repair and alteration work may be accomplished by a mechanic, a repair man, or a person (perhaps yourself) working *under the supervision* of a mechanic or repair man providing the supervisory mechanic personally observes the work being done to the extent necessary to insure that it is accomplished properly. The supervisor must be readily available, in person, for consultation during the process. Major repairs include the overhaul of a power plant, the overhaul of a propeller, the calibration and repair of instruments, etc.

Engine Overhaul

As engines begin to approach their overhaul life limit, they should be watched for initial signs of malfunctioning in order to prevent failure. The following lists typical indicators of a need for an overhaul.

■ Has there been any unusual increase in oil consumption over a 25 hour period? Similarly, has fuel consumption increased or drifted to a new mixture setting?

■ What is the general health history of the engine? What kind of maintenance has the engine received?

■ What has the trend been in differential compression checks? What story do the spark plugs tell?

Table 7-1. Recommended Overhaul Periods for Avco Lycoming Engines Based on continuous Service (20 to 30 Hours per Month). (Reprinted from Avco Lycoming Service Instruction No. 1009T, Courtesy of Avco Lycoming).

RECOMMENDED OVERHAUL PERIODS

FIXED WING AIRCRAFT

Engine Models	Hours
O-235, C, -D, -F, -G, O-290-D	2000
O-290-D2	1500
O-320; IO-320-A, -E	2000
IO-320-B, -D	2000
IO-320-C	1500
AIO-320 (160 hp)	1000
O-340 Series	2000
O-360; IO-360-B, -E, -F (180 hp)	2000
IO-360-A, -C, -D Series (200 hp)	1600
AIO-360 (200 hp)	1000
TIO-360	1200
O-435; GO-435	1200
GO; GSO-480; IGSO-480-A1E6, -A1G6	1400
IGSO-480-A1A6, -A1B6 -A1C6, -A1F6	1400
O-540-B (235 hp)	1800
O-540-A, IO-540-C (250 hp)	1800
O-540-E, -F, -G, -H;IO-540-D, -N, -R (260 hp)'	1800
IO-540-A, -B (290 hp)	1200
IO-540-E, -G, -P (290 hp)	1400
IO-540-J	1500
IO-540-K, -M (300 hp)	1800
IGO-540-B; IGSO-540-B	1200
IGO-540-A; IGSO-540-A	1200
TIO-540-A, -C	1800
TIO-540-F	1500
TIO-540-J	1200
TIO-541-A (310 hp)	1300
TIO-541-E (380 hp)	1200
TIGO-541-E	1200
IO-720 Series	1500

ROTARY WING AIRCRAFT

Engine Models	Hours
O-360-C2B, -C2D	1200
HO-360; HIO-360-B	
HIO-360-A, -C, -D Series	1000
VO-360-A Series	600
VO-360-B; IVO-360	1000
VO-435-A Series	1200
VO-435-B Series	1200
TVO-435 Series	1000
VO-540 Series	1200
IVO-540 Series	600
TVO, TIVO-540 Series	1200

■ Has there been evidence of metal chips in the oil screen? Any time metal in the amount of one-half teaspoon or more is collected in the screen, it is usually grounds for engine overhaul.

■ Are the engine temperatures still running normal? Does the engine still develop rated static RPM?

■ Most important of all, does the pilot have confidence in the engine?

In the event the answer to any of these questions is unsatisfactory, a basis exists for engine overhaul, remanufacture or replacement. The terms *overhaul* and *remanufacture* are often used interchangeably but in reality they mean different things. To "remanufacture" an engine is to restore the engine and accessories to new engine tolerance limits. An overhauled engine can be restored to factory specified service limit tolerances. In some instances, a service limit tolerance might be as much as twice that of a new engine tolerance for the same part. An engine manufacturer can "zero time" an engine by the remanufacture process. Independent shops often remanufacture engines by working to new engine tolerances. However, they may not zero time an engine. The total engine time must appear in the logbook.

After refurbishment, an engine is test run to verify performance as well as supply the user an engine which can be operated at full power in an aircraft. Table 7-2 lists a typical run-in acceptance test for a Continental IO-520 engine.

A significant difference exists in the cost of restoring an engine to service depending upon the process used. Typically, a Continental 0-200 might cost $1400 to overhaul, $1600 to $2000 to exchange-remanufacture, $2500 to exchange-new or $3200 new (no exchange). The T.W. Smith Company of Cincinnati, Ohio, a major independent remanufacturer of engines lists engine exchange prices as shown in Table 7-3 (typical 1974 prices). An exchange program is one way to minimize down time.

Parts Inspection

A major part of an engine overhaul or remanufacture, perhaps 20 to 40 percent, is devoted to the inspection of parts. Inspection techniques include visual, magnaflux, magaglow, die-penetrant and radiographic (X-ray).

Magnetic particle inspection (*magnaflux* or *magnaglow*) is a method of detecting invisible cracks and other such defects in ferrous materials such as iron and steel. The technique is non-destructive and is particularly valuable as a means of examining highly stressed engine parts for fatigue defects (Fig. 7-1).

Period	Time-Minutes	RPM	
1	15	2400	Warm-Up
2	10	2600	
3	10	2700	
4	5	2850-2900	100% Power
5	5	2400	60% Power
6	5	600 ± 25	Cool Down Period

Stop engine, drain oil, clean screen (replace filter).
weigh oil in for oil consumption determination.

START OIL CONSUMPTION DETERMINATION

Period	Time-Minutes	RPM	
7	5	2400	Warm-Up
8	10	2550 *	
9	10	2700	
10	10	2850-2900	Full Throttle
		2100 **	Check Magnetos
11	5	600 ± 25	Cool Down Period

* *Readings must be recorded after completion of each 10 minute interval during oil consumption run.*

** *Magneto spread to be taken after completion of oil consumption run. Engine must be throttled to specified RPM and temperature allowed to settle out before taking magneto spread.*

*** *Stop engine, drain oil, weigh oil and record engine oil consumption. Oil consumption at a rate of 1.80 lbs./ - 1/2 hr. maximum is acceptable. If value in excess of 1.80 lbs./ - 1/2 hr. is determined, re-run 1 hour. If consumption is still excessive, engine must be completely rechecked for construction.*

Table 7-2. Standard Acceptance Test for the Continental 10-520 300 hp Engine After Overhaul (courtesy of Teledyne Continental Motors).

CONTINENTAL

ENGINE MODEL	H. P. RATING	EXCHANGE PRICE
C90-8F, 12F, 14F, 16F	90	$1,545.00
0-200-A (Delco)	100	1,545.00
0-200-A (Prestolite)	100	1,645.00
0-200-B	100	1,645.00
0-300-A, B	145	1,905.00
0-300-C, D	145	2,005.00
GO-300-A, C	175	2,655.00
GO-300-D, E	175	2,805.00
E185-11	205	2,755.00
E225-4, 8	225	2,855.00
0-470-J	225	2,855.00
0-470-K, L, R	230	2,855.00
0-470-G	240	2,855.00
0-470-M	240	3,255.00
IO-346-A	210	2,555.00
IO-360-A	210	3,055.00
IO-360-C, D, G, H	210	3,155.00
IO-470-C,	250	3,455.00
IO-470-D, E, F, M	260	3,505.00
IO-470-J, K	225	3,505.00
IO-470-H, L, S	260	3,555.00
IO-470-N, U, V	260	3,595.00
IO-520-A, B, BA, C, D, E, F, K, L	285/300	3,795.00
TSIO-360-A, B	210	3,555.00
TSIO-360-C, D	260	3,655.00
TSIO-470 B, C, D	260	4,095.00
TSIO-520-B, G	285	4,305.00
TSIO-520-C	285	4,495.00

LYCOMING

ENGINE MODEL	H. P. RATING	EXCHANGE PRICE
0-235-CI, CIB	115	$1,685.00
0-290-D-D2	130/140	1,985.00
0-320 Series	150/160	2,085.00
0-340	170	2,355.00
IO-320-BIA	160	2,435.00
IO-320-CIA	160	2,555.00
LIO-320-BIA	160	2,495.00
0-360 Series	168-180	2,355.00
IO-360-AIA, AID, CIC, CIE	200	2,795.00
IO-360-BIA, BIB, BIE	180	2,635.00
0-540 Series	235/250/260	3,155.00
IO-540-A, B, E	290	3,895.00
IO-540-C, D	250/260	3,355.00
IO-540-G	290	3,795.00
IO-540-J	250	3,495.00
IO-540-K, M	300	4,395.00
TIO-540-A	310	6,155.00
TIO-541-A	310	7,195.00
TIO-541-E	380	7,395.00
IO-720-A	400	6,195.00
GO-435-C2B	260	4,405.00
GO-480-B, F, D, G	270/275/295	4,675.00
GO-480-G1B6	295	5,495.00
GSO-480-B	340	5,655.00
IGSO-480-A	340	6,205.00
IGO-540-A	350	5,895.00
IGSO-540-A	380	7,605.00
IGSO-540-B	380	7,675.00

Table 7-3. A Partial Listing of 1974 Engine Prices (courtesy of T.W. Smith Aviation Inc.).

Fig. 7-3. A dynamic duo, the Hirth two-cylinder, two-cycle engine that powers the Bede 5 aircraft develops 70hp at 6500RPM (courtesy of Bede Aircraft).

For take-off, use minimum power to attain at least 40 MPH IAS before applying required take-off power.

Do not use take-off power any longer than necessary.

Use minimum rate of climb with maximum air speed.

Do not lug engine. Use low pitch range for propeller setting with the 75 percent manifold pressure required for clean in-flight attitude. For aircraft not equipped with variable pitch propeller or manifold pressure gauge, use 75 percent power for clean in-flight attitude.

After desired altitude is reached, maintain level flight attitude. At some period between 20 and 30 minutes, the cylinder head temperature will show a rapid decrease. This indicates that the piston rings have seated and the short ground runs and the in-flight engine break-in procedures were satsifactorily accomplished.

Return to airport and weigh or measure amount of lubricating oil. At the same time, note oil temperature and ground attitude of aircraft. This applies to oil stick measurement only.

Again fly aircraft at 75 percent cruise power setting for one hour, land, and again weigh or measure oil. If oil is measured on stick gauge, make sure temperature and aircraft ground attitude are the same as described in the preceding paragraph.

Record test flight and oil consumption in engine log book before aircraft is released.

After release of aircraft to operator, discourage prolonged ground runs. This can be highly detrimental to any engine."

For years Lycoming has offered new engines with chrome plated cylinder barrels as an option. Continental also offers chrome plated cylinders, but, currently only on their remanufactured and overhauled engines. Considering the 1000-hour/3-year warranty offered by Chrome Plate Inc. and the fact that refurbished chrome cylinders are about half the cost of new ones—well now, that's something to consider.

Engine Reliability

The modernlightplane engine is a genuine tribute to reliability. From a time when overhauls were accomplished after 15 hours of use during World War I, to the Lindbergh era where 300 hours represented the life of an engine, to our modern 2000-hour time-between-overhaul powerplant is indeed an accomplishment. Improved materials, production of parts to close tolerances,

Fig. 7-4. Lightplane engine of the future? The Sermel TRS-18 jet is certainly a first. This unique powerplant develops an equivalent 173hp at a cruise airspeed of 325MPH (courtesy of Bede Aircraft).

established maintenance procedures, airworthiness directives—all have contributed to the engine reliability we enjoy today. Engine failure, once a major cause of accidents, is slowly being relegated to a minor causitive factor.

A recent special study, "Accidents Involving Engine Failure/Malfunction, U.S. General Aviation, 1965-1969" found there were 4,310 fixed wing general aviation airplanes involved in accidents that were precipitated by engine failure or malfunction. Of these, 841 aircraft were destroyed and the remainder were damaged substantially. The pilot in command was cited as the probable cause in 51.8 percent of the engine-failure accidents. The powerplant itself was cited as a probably cause in 44.62 percent of the cases.

Particularly frequent cause factors were found to be as follows:

Pilot-in-Command Involvement:

Inadequate preflight preparation or planning.....934 instances
Mismanagement of fuel.............................615 instances
Improper operation of powerplant/its controls ..504 instances
Improper in-flight decisions or planning ....127 instances
Became lost or disoriented..........................101 instances

Powerplant Involvement:

Valve assemblies ....130 instances
Carburetor..102 instances
Master and connecting rods ....86 instances
Cylinder assembly ....72 instances
Piston and rings......................................70 instances
Magnetos..64 instances
Crankshaft..57 instances
Sparkplug...53 instances

Maintenance, servicing and inspection personnel were cited as a probable cause/related factor in 425 (9.9 percent of the engine-failure accidents. The predominent cause/factor citations were:

—Inadequate mainteance and inspection.
—Improper maintenance (maintenance personnel).
—Improperly serviced aircraft (owner-pilot).
—Improper maintenance (owner personnel).

As engine manufacturers continue to work on improved valve assemblies, carburetors, etc., perhaps we can participate by reducing that 9.9 percent figure attributable to inadequate maintenance.

Table 7-4. The Teledyne Continental Family of Opposed Aircraft Engines.

MODEL	NO. of CYL.	CYLINDER ARRANGE-MENT	METO POWER HP @ RPM	BORE & STROKE	DISP. CU/IN
C-90 Series	4	HO	90-2475	4.06 x 3.87	201
0-200-A	4	HO	100-2750	4.06 x 3.87	201
0-200-B	4	HO	100-2750	4.06 x 3.87	201
0-300-A, B & C	6	HO	145-2700	4.06 x 3.87	301
0-300-D & E	6	HO	145-2700	4.06 x 3.87	301
IO-346-Series	4	HO	165-2700	5.25 x 4	346
IO-360-A, C & D	6	HO	210-2800	4.44 x 3.87	360
IO-360-B	6	HO	180-2700	4.44 x 3.87	360
**TSIO-360-A & B	6	HO	210-2800	4.44 x 3.87	360
**TSIO-360-C	6	HO	225-2800	4.44 x 3.87	360
0-470-J	6	HO	225-2550	5 x 4	471
0-470-R	6	HO	230-2600	5 x 4	470
0-470-4, 11 & 13	6	HO	225-2600	5 x 4	471
0-470-15	6	HO	190-2300	5 x 4	471
IO-470-C	6	HO	250-2600	5 x 4	471
IO-470-D	6	HO	260-2625	5 x 4	471
IO-470-E	6	HO	260-2625	5 x 4	471
IO-470-F	6	HO	260-2625	5 x 4	471
IO-470-G	6	HO	250-2600	5 x 4	471
IO-470-H	6	HO	260-2625	5 x 4	471
IO-470-J	6	HO	225-2600	5 x 4	471
IO-470-K	6	HO	225-2600	5 x 4	471
IO-470-L	6	HO	260-2625	5 x 4	471
IO-470-M	6	HO	260-2625	5 x 4	471
IO-470-N	6	HO	260-2625	5 x 4	471
IO-470-S	6	HO	260-2625	5 x 4	471
IO-470-U	6	HO	260-2625	5 x 4	471
IO-470-V	6	HO	260-2625	5 x 4	471
**TSIO-470-B, C & D	6	HO	260-2600	5 x 4	471
IO-520-A & J	6	HO	285-2700	5.25 x 4	520
IO-520-B	6	HO	285-2700	5.25 x 4	520
IO-520-C	6	HO	285-2700	5.25 x 4	520
IO-520-D	6	HO	285-2700	5.25 x 4	520
IO-520-E	6	HO	285-2700	5.25 x 4	520
IO-520-F	6	HO	285-2700	5.25 x 4	520
IO-520-K	6	HO	285-2700	5.25 x 4	520
IO-520-L	6	HO	285-2700	5.25 x 4	520
**TSIO-520-B	6	HO	285-2700	5.25 x 4	520
**TSIO-520-C & H	6	HO	285-2700	5.25 x 4	520
**TSIO-520-D	6	HO	285-2700	5.25 x 4	520
**TSIO-520-E	6	HO	300-2700	5.25 x 4	520
**TSIO-520-G	6	HO	285-2600	5.25 x 4	520
**TSIO-520-J	6	HO	310-2700	5.25 x 4	520
**TSIO-520-K	6	HO	285-2700	5.25 x 4	520
**GTSIO-520-C	6	HO	340-3200	5.25 x 4	520
**GTSIO-520-D	6	HO	375-3400	5.25 x 4	520
**GTSIO-520-F	6	HO	435-3400	5.25 x 4	520
**GTSIO-520-G	6	HO	375-3400	5.25 x 4	520
**GTSIO-520-H	6	HO	375-3400	5.25 x 4	520
Tiara 6-285	6	HO	285-4000	4.88 x 3.62	406
Tiara 6-320	6	HO	320-4400	4.88 x 3.62	406

*Less Turbocharger **Turbocharged

ENGINE DIMENSIONS			WEIGHT DRY LBS. BASIC ENG.	PROP. DRIVE	FUEL GRADE	COMP. RATIO
LENGTH	WIDTH	HEIGHT				
31.25	31.50	24.19	186	Direct	80/87	7.0:1
28.50	31.56	23.18	190	Direct	80/87	7.0:1
28.50	31.56	23.18	220	Direct	80/87	7.0:1
39.75	31.50	23.25	268	Direct	80/87	7.0:1
36.00	31.50	27.00	268	Direct	80/87	7.0:1
30.00	33.38	22.40	297	Direct	91/96	7.5:1
34.60	31.40	24.33	327	Direct	100/130	8.5:1
36.03	31.40	22.34	334	Direct	80/87	6.5:1
*35.84	33.03	23.75	334	Direct	100/130	7.5:1
*35.84	33.03	23.75	300	Direct	100/130	7.5:1
36.03	33.32	27.75	381	Direct	80/87	7.0:1
36.03	33.56	28.42	438	Direct	80/87	7.0:1
40.90	33.64	26.44	414	Direct	80/87	7.0:1
44.75	33.40	30.10	405	Direct	91/96	7.0:1
37.93	33.58	26.81	432	Direct	91/96	8.0:1
43.31	33.56	19.75	426	Direct	100/130	8.6:1
42.76	33.56	19.75	461	Direct	100/130	8.6:1
37.22	33.56	23.79	464	Direct	100/130	8.6:1
42.76	33.56	19.75	461	Direct	100/130	8.0:1
38.14	33.58	26.81	457	Direct	100/130	8.6:1
38.14	33.39	26.81	413	Direct	80/87	7.0:1
38.14	33.39	26.81	413	Direct	80/87	7.0:1
43.17	33.56	19.75	473	Direct	100/130	8.6:1
47.16	33.56	19.75	474	Direct	100/130	8.6:1
38.14	33.58	26.81	457	Direct	100/130	8.6:1
41.41	33.56	19.75	464	Direct	100/130	8.6:1
44.14	33.86	19.75	471	Direct	100/130	8.6:1
43.69	33.56	19.75	472	Direct	100/130	8.6:1
*39.52	33.56	20.25	511	Direct	100/130	7.5:1
41.41	33.56	19.75	471	Direct	100/130	8.5:1
39.71	33.58	26.71	457	Direct	100/130	8.5:1
42.88	33.56	19.75	449	Direct	100/130	8.5:1
37.36	35.46	23.79	454	Direct	100/130	8.5:1
47.66	33.56	19.75	475	Direct	100/130	8.5:1
41.41	35.91	19.75	455	Direct	100/130	8.5:1
40.91	33.56	19.75	466	Direct	100/130	8.5:1
40.91	33.56	23.25	466	Direct	100/130	8.5:1
*39.75	33.56	20.32	483	Direct	100/130	7.5:1
*40.91	33.56	20.04	460	Direct	100/130	7.5:1
*43.25	33.58	22.34	484	Direct	100/130	7.5:1
*39.75	33.56	20.32	483	Direct	100/130	7.5:1
*40.91	33.56	20.04	459	Direct	100/130	7.5:1
54.36	33.56	22.50	486	Direct	100/130	7.5:1
54.36	33.56	20.32	452	Direct	100/130	7.5:1
*42.56	34.04	23.10	557	Geared	100/130	7.5:1
*42.56	34.04	26.78	550	Geared	100/130	7.5:1
56.25	34.04	26.18	600	Geared	100/130	7.5:1
*42.56	34.04	26.78	557	Geared	100/130	7.5:1
*42.56	34.04	26.78	550	Geared	100/130	7.5:1
40.11	32.91	24.22	409	Geared	100/130	9.0:1
40.11	32.91	24.22	409	Geared	100/130	9.5:1

Table 7-5. TRS-18 Specifications.

Specification	Value
Maximum continuous thrust	200 lb.
Corresponding fuel consumption	283 lb./hr.
Fuel consumption at cruise	102 lb./hr.
Electric generator	600W at 28V DC
Weight	66 lb.
Overall length	22.6 in.
Maximum running altitude	40,000 ft.
Fuel types	JPI-JPA-JP5
Oil types	Standard Turbine Jet Oil

On the Horizon

The future for lightplane engines continues to be filled with many bright stars. Continental, in their Tiara series, is bringing forth a new family of lightweight geared powerplants designed for economy in production as well as maintenance (See Table 7-4). The Tiara program will foster the development of low RPM propellers and as a result will reduce noise and pilot fatigue.

On the ultralightweight scene, the 43.9-cubic-inch, 2-cylinder, 2-cycle horsepower Hirth engine of 70 horsepower that powers the BD-5 aircraft is certainly a new image in aircraft engines. Ultrasmall, ultralight, and very powerful at 1.17 lb./hp., the engine at idle sounds like a mad bee and at cruise it sounds like the entire hive.

Unique among future lightplane engines, the TRS-18 turbojet engine used in the BD-5J, a personal jet aircraft, is a recent development of the Sermel Co. of France. The engine is manufactured in the U.S. by the Ames Industrial Corporation. The engine features automatic monitoring equipment. Included is automatic fuel shut off. Starting is completely automatic and does not require auxilliary power. The TRS-18 is a compact turbojet engine with its main accessories enclosed in a cylindrical housing 12.5″ in diameter. It consists of a single-stage, single-axial entry centrifugal compressor and a single-stage axial flow turbine wheel. The compressor and turbine rotation elements are mounted on a single shaft supported by two bearings. A gearbox drives the different accessories. The lubrication system is completely self-contained in the basic package. Principle specifications for the TRS-18 are shown in Table 7-5.

Will the day come when we will jet about the skies and write about modern lightplane jet engines? Progress being what it is, we well might. As Samual Butler once stated, "All progress is based upon universal innate desire on the part of every organism to live beyond its income"; truly the words of an aviation prophet. So . . . see you at flight level 240, about 1985.

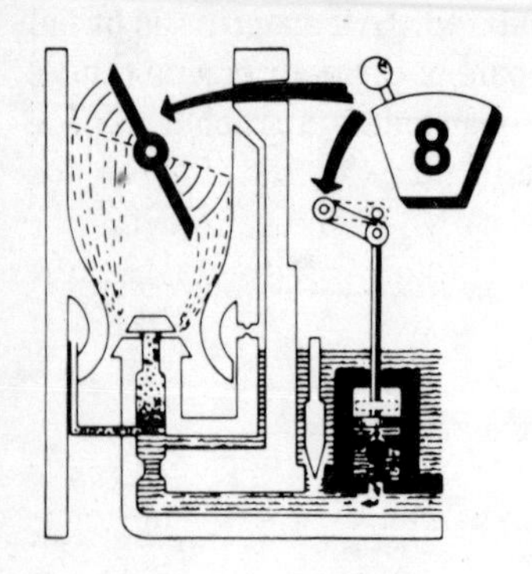

Aircraft Engine Overhaul Standards: FAA Advisory Circular AC 43.13-1A And 2

All moving or highly stressed parts and those subjected to high temperature should have a critical visual inspection at the time of overhaul. It is often necessary to supplement the visual inspection by employing one of the following procedures:

a. Wet or dry magnetic dust inspection of magnetic material,

b. Wet or dry penetrant inspection of non-magnetic materials,

c. X-ray or sonic inspection of any material, or

d. Hydrostatic inspection testing of fluid lines and internal passages and assemblies such as cylinder heads.

POWERPLANT SUDDEN STOPPAGE

For the purpose of this section, powerplant sudden stoppage refers to any momentary slowdown or complete stoppage of the main shaft of an aircraft powerplant, when the stoppage of a reciprocating or turboprop engine is the result of the rotating propeller striking a foreign object or when the stoppage of a turbine engine is the result of ingestion of foreign objects or material. Any aircraft powerplant that has been subjected to sudden stoppage should be inspected to the extent necessary to assure continued safe operation. These procedures will serve as a guide for locating damage that might occur whenever an aircraft powerplant has been subjected to sudden deceleration or stoppage.

To fully evaluate any unsatisfactory findings resulting from this type of inspection, it will be necessary to refer to the applicable manufacturer's service and overhaul data. In addition, many of the prime aeronautical engine manufacturers now have specific recommendations on the subject of sudden stoppage involving their products. To assure continued airworthiness and reliability, it is essential that such data be used. In the event the manufacturer has not specified an instruction to follow, the following can be used as a guideline.

a. Reciprocating Engine (Direct Drive)

(1) Powerplant Exterior Inspection. Remove the engine cowling and examine the engine for visible external damage and audible internal damage.

(a) Rotate the propeller shaft to determine any evidence of abnormal grinding or sounds.

(b) With the propeller removed, inspect the crankshaft flange or splines for signs of twisting, cracks or other deformation. Remove the thrust-bearing nut and seal and thoroughly inspect the threaded area of the shaft for evidence of cracks.

(c) Rotate the shaft slowly in 90° increments while using a dial indicator or an equivalent instrument to check the concentricity of the shaft.

(d) Remove the oil sump drain plug and check for metal chips and foreign material.

(e) Remove and inspect the oil screens for metal particles and contamination.

(f) Visually inspect engine case exterior for signs of oil leakage and cracks. Give particular attention to the propeller thrust-bearing area of the nose case section.

(g) Inspect cylinders and cylinder holddown area for cracks and oil leaks. Thoroughly investigate any indication of cracks, oil leaks or other damage.

(2) Powerplant Internal Inspection

(a) On engines equipped with crankshaft vibration dampers, remove the cylinders necessary to inspect the dampers and inspect in accordance with the engine manufacturer's inspection and overhaul manual. When engine design permits, remove the damper pins and examine the pins and damper liners for signs of nicks or brinelling.

(b) After removing the engine-driven accessories, remove the accessory drive case and examine the accessory and supercharger drive gear train, couplings, and drive case for evidence of damage.

1. Check for cracks in the case in the area of accessory mount pads and gear shaft bosses.

2. Check the gear train for signs of cracked, broken or brinelled teeth.

3. Check the accessory drive shaft couplings for twisted splines, misalignment and runout.

(3) Accessory and Drive Inspection

Check the drive shaft of each accessory, i.e., magnetos, generators, external supercharger and pumps for evidence of damage.

(4) Engine Mount Inspection

(a) Examine the engine flex mounts when applicable for looseness, distortion, or signs of tear.

(b) Inspect the engine mount structure for bent, cracked or buckled tubes.

(c) Check the adjacent airframe structure for cracks, distortion, or wrinkles.

(d) Remove engine mount bolts and mount holddown bolts and inspect for shear, cracks or distortion.

(5) Exhaust Driven Supercharger (Turbo) Inspection

Sudden stoppage of the powerplant can cause the heat in turbine parts to heat soak the turbine seals and bearings. This excessive heat causes carbon to develop in the seal area and varnish to form on the turbine bearings and journals.

(a) Inspect all air ducts and connections for air leaks, warpage or cracks.

(b) Remove compressor housing and check the turbine wheel for rubbing or binding.

(6) Propeller Inspection Repair

Any propeller that has struck a foreign object during service should be promptly inspected in accordance with the propeller manufacturer's prescribed procedures for possible damage resulting from this contact with the foreign object and, if necessary, repair according to the manufacturer's instructions. If the propeller is damaged beyond the repair limits established by the propeller manufacturer and a replacement is necessary, install the same make/model or alternate approved for this installation. Refer to the aircraft manufacturer's optional equipment list, applicable FAA Aircraft Specification, Data Sheet or Supplemental Type Certificate Data.

b. Reciprocating Engine (Geared Drive). Inspect the engine, propeller and components as in preceding paragraphs.

(1) Remove the propeller reduction gear housing and inspect for:

(a) Loose, sheared or spalled cap screws or bolts.

(b) Cracks in the case.

(2) Disassemble the gear train and inspect the propeller shaft, reduction gears and accessory drive gears for nicks, cracks or spalling.

c. Turbine, Engine, Ingestion Inspection. When the components of the compressor assembly or turbine section are subjected to ingestion damage, refer to the engine manufacturer's inspection and overhaul manual for specific inspection procedures and allowable tolerances. In general, an inspection after ingestion of foreign materials consists of the following areas:

(1) Inspect the external areas of the engine cases, attached parts and engine mounts for cracks, distortion or other damage.

(2) Inspect the turbine disc for warpage or distortion.

(3) Inspect turbine disc seal for damage from rubbing or improper clearance.

(4) Inspect compressor rotor blades and stators for nicks, cracks or distortion.

(5) Check rotor and main shaft for misalignment.

(6) Inspect shaft bearing area for oil leaks.

(7) Inspect the hot section for cracks or hot spots.

(8) Inspect the accessory drives as prescribed under paragraph a (2) (b).

NOTE: Turbine disc seal rubbing is not unusual and might be a normal condition. Consult the engine manufacturer's inspection prodecures and table of limits.

d. Turboprop Engine Inspection

(1) When sudden stoppage is the result of compressor ingestion of foreign material, inspect the engine as follows:

(a) Inspect the powerplant as described in paragraph c, "Turbine Engine, Ingestion Inspection."

(b) Inspect the reduction gear section as described in paragraph b(1) and b(2) where reduction gear damage is suspected.

(2) When sudden stoppage is the result of the propeller striking foreign objects, inspect the engine as follows:

(a) Inspect the reduction gear section as described in paragraph b(1) and b(2).

(b) Inspect mainline shafts and coupling shafts for runout and spiral cracks.

(c) Inspect bearings for brinelling.

(d) Inspect engine compressor and turbine blades for tip clearance.

e. Approval for Returning Engine to Service

(1) Correct all discrepancies found in the foregoing inspection in accordance with the engine manufacturer's service instructions.

(2) Test run the engine to determine that the engine, propeller and accessories are functioning properly.

(3) After shutdown, check the engine for oil leak and check oil screens for signs of contaminants.

(4) If everything is normal the engine is ready for preflight runup and test flight.

CRANKSHAFTS

Carefully inspect for misalignment and replace if bent beyond the manufacturer's permissible service limit. Make no attempt to straighten crankshafts damaged in service without consulting the engine manufacturer for appropriate instructions. Worn journals may be repaired by regrinding in accordance with manufacturers' instructions. It is recommended that grinding operations be performed by appropriately rated repair stations or the original engine manufacturer to assure adherence to aeronautical standards.

Common errors that occur in crankshaft grinding are the removal of nitrided journal surface, improper journal radii, unsatisfactory surfaces and grinding tool marks on the journals. If the fillets are altered, do not reduce their radii. Polish the reworked surfaces to assure removal of all tool marks. Most opposed engines have nitrided crankshafts and manufacturers specify that these crandshafts must be nitrided after grinding.

REPLACEMENT PARTS IN CERTIFICATED ENGINES

Only those parts which are approved under FAR Part 21 should be used. Serviceable parts obtained from the engine manufacturer, his authorized service facility, and those which are FAA/PMA approved meet the requirements of FAR Part 21 and are acceptable for use as replacement parts. It is suggested that the latest type parts as reflected on the current parts list be obtained. Parts from military surplus stocks, which are applicable to the specific engine may be used provided they were originally accepted under the military procurement agency's standards, are found to be serviceable and are not prohibited from use by the Administrator.

a. Parts For Obsolete Engines. Parts that are no longer obtainable from the original manufacturer or his successor are sometimes fabricated locally. When it becomes necessary to do this, physical tests and careful measurements of the old part may provide adequate technical information. This procedure is usually regarded as a major change which requires engine testing and is not recommended except as a last alternative.

b. Military Surplus. Surplus tank parts or ground power unit engines are used on engines used in restricted aircraft and amateur-built aircraft. Users of such parts are cautioned to determine that they do not exceed the design limits of the engine. For example, a particular tank engine utilized a piston design that developed a compression ratio well in excess of the crankshaft absorption rate of the aircraft engine counterpart—result, crankshaft failure.

CYLINDER HOLDDOWN NUTS AND CAPSCREWS

Great care is required in tightening cylinder holddown nuts and capscrews. They must be tightened to recommended torque limits to prevent improper stressing and to insure even loading on the cylinder flange. The installation of baffles, brackets, clips, and other extraneous parts under nuts and capscrews is not considered good practice and should be discouraged.

If these baffles, brackets, etc., are not properly fabricated or made of suitable material, they cause loosening of the nuts or capscrews even though the nuts or capscrews were properly tightened and locked at installation. Either improper prestressing or loosening of any one of these nuts or capscrews will introduce the danger of progressive stud failure with the possible loss of the engine cylinder in flight. Do not install parts made from aluminum alloy or other soft metals under cylinder holddown nuts or capscrews.

REUSE OF SAFETY DEVICES

Do not use cotter pins and safety wire a second time. Flat steel-type wristpin retainers and thin lockwashers likewise should be replaced at overhaul unless the manufacturer's recommendations permit their reuse.

SELF-LOCKING NUTS FOR AIRCRAFT ENGINES AND ACCESSORIES

Self-locking nuts may be used on aircraft engines provided the following criteria are met:

a. Where their use is specified by the engine manufacturer in his assembly drawing, parts list, and bills of material.

b. When the nuts will not fall inside the engine should they loosen and come off.

c. When there is at least one full thread protruding beyond the nut.

d. When the edges of cotter pin or lockwire holes are well rounded to preclude damage to the locknut.

e. Prior to reuse, the effectiveness of the self-locking feature is found to be satisfactory.

f. Where the temperature will not exceed the maximum limits established for the self-locking material used in the nut. On many engines the cylinder baffles, rocker box covers, drive covers and pads, and accessory and supercharger housings are fastened with fiber insert locknuts which are limited to a maximum temperature of 250 F. Above this temperature, the fiber insert will usually char and consequently lose its locking characteristic. For locations such as the exhaust pipe attachment to the cylinder, a locknut which has good locking features at elevated temperatures will give invaluable service. In a few instances, fiber insert locknuts have been approved for use on cylinder holddown studs. This practice is not generally recommended. Especially tight stud fits to the crankcase must be provided and extremely good cooling must prevail so that low temperatures exist at this location on the specific engine for which such use is approved.

g. All proposed applications of new types locknuts or new applications of currently used self-locking nuts must be investigated because many engines require specifically designed nuts. Such specifically designed nuts are usually required for one or more of the following reasons to provide:

(1) Heat resistance.

(2) Adequate clearance for installation and removal.

(3) For the required degree of tightening or locking ability which sometimes requires a stronger, specifically heat-treated material, a heavier cross-section or a special locking means.

(4) Ample bearing area under the nut to reduce unit loading on softer metals.

(5) To prevent loosening of studs when nuts are removed.

h. Information concerning approved self-locking nuts and their use on specific engines is usually found in engine manufacturer's manuals or bulletins. If the desired information is not available, it is suggested that the engine manufacturer be contacted.

WELDING OF HIGHLY STRESSED ENGINE PARTS

In general, welding of highly stressed engine parts is not recommended for parts that were not originally welded. However, under the conditions given below, welding may be accomplished if it can be reasonably expected that the welded repair will not adversely affect the airworthiness of the engine when:

a. The weld is externally situated and can be inspected easily.

b. The part has been cracked or broken as the result of unusual loads not encountered in normal operation.

c. A new replacement part of obsolete type engine is not available.

d. The welder's experience and equipment employed will insure a first-quality weld in the type of material to be repaired and will insure restoration of the original heat treat in heat-treated parts.

WELDING OF MINOR ENGINE PARTS

Many minor parts not subjected to high stresses may be safely repaired by welding. Mounting lugs, cowl lugs, cylinder fins, covers, and many parts originally fabricated by welding are in this category. The welded part should be suitably stress-relieved after welding.

METALLIZING

Metallizing internal parts of aircraft engines is not acceptable unless it is proven to the Federal Aviation Administration that the metallized part will not adversely affect the airworthiness of the engine. Metallizing the finned surfaces of steel cylinder barrels with aluminum is acceptable. Many engines are originally manufactured in this manner.

PLATING

Restore plating on engine parts in accordance with the manufacturer's instructions. In general, chromium plating should not be applied to highly stressed engine parts. Certain applications of this nature have been found to be satisfactory. However, engineering evaluation of the details for the processes used should be obtained.

a. Dense Chromium Plating. Plating of the crankpin and main journals of some small engine crankshafts has been found satisfactory, except where the crankshaft is already marginal in strength. Plating to restore worn low-stress engine parts such as accessory driven shafts and splines, propeller shaft ends, and seating surfaces of roller- and ball-type bearing races is acceptable.

b. Porous Chromium Plated. Plating the walls of cylinder barrels have been found to be satisfactory for practically all types of engines. Dense or smooth chromium plating without roughened surfaces, on the other hand, has not been found to be generally satisfactory.

(1) Cylinder barrel pregrinding and chromium plating techniques used by the military are considered acceptable for all engines. Military approved facilities engaged in doing this work in accordance with military specifications are eligible for approval by the Federal Aviation Administration.

(2) Chromium plated cylinder barrels have been required for some time to be indentified in such a manner that the markings are visible with the cylinder installed. Military processed cylinders are banded with orange enamel above the mounting flange. It has been the practice to etch on either the flange edge or on the barrel skirt the processer's initials and cylinder oversize. Most plating facilities use the orange band as well as the permanent identification marks.

Table 8-1. Engine and Maximum Cylinder Barrel Oversize.

Engine manufacturer	Engine series	Maximum oversize (in).
Air Cooled Motors (Franklin).	No oversize for sleeved cylinders.	
	Solid cylinders	0.020
Continental Motors	R-670, W-670, R9A	0.020
	All others	0.015
Jacobs	All	0.020
Kinner	All	0.015
Pigman, LeBlond, Rearwin, Ken Royce.	All	0.025
Lycoming	All	0.015
Menasco	All	0.010
Pratt & Whitney	R-2800B, C, CA, CB	0.025
	*R-985 and R-1830	0.030
	All others	0.020
Ranger	6-410 early cyls. 6-390	0.010
	6-410 late cyls. 6-440 (L-440) series.	0.020
Warner	All	0.015
Wright	All	0.020

*(The above oversize limits correspond to the manufacturer's requirements, except for P&W R-985 and R-1830 series engines.)

(3) A current list of engine and maximum permissible cylinder barrel oversize is shown in Table 8-1.

(4) The following is a list of known agencies, and their identifying initials, performing cylinder barrel plating for the military service.

Agencies	Initials
Hol-Chrome Corporation	HCC
San Antonio Air Materiel Area	SAX
Koppers Co. (American Hammered Piston Ring)	KC
McQuery-Norris	MQN
VanDer Horst Corporation	VHD
Terry Industries	TIX
Lement Chromium	LC
Pennington Channel Chrome	PCC
Spar-Tan Engineering Company	SEC
Superior Aero Chrome	SAC

(5) Cylinder barrels which have been plated by an agency using a process approved by the FAA and which have not been preground beyond maximum permissible limits will be considered acceptable for installation on certificated engines. It will be the responsibility of the owner or the repairing agency to provide this proof. In some cases, it might be necessary to remove cylinders to determine the amount of oversize because this information might be etched on the mating surface of the cylinder base flange.

ENGINE ACCESSORIES

Overhaul and repair of engine accessories in accord with the engine and the accessory manufacturers' instructions are recommended.

CORROSION

Accomplish corrosion preventive measures for temporary and dead storage in accord with the instructions issued by the pertinent engine manufacturer. Avoid the use of strong solutions which contain strong caustic compounds and all solutions, polishes, cleaners, abrasives, etc., which might possibly promote corrosive action.

ENGINE RUN-IN

After an aircraft engine has been overhauled, it is recommended that the pertinent aircraft engine manufacturer's run-in instructions be followed. Observe the manufacturer's recommendations concerning engine temperatures and other criteria.

Repair processes employed during overhaul often necessitate amending the manufacturer's run-in procedures. Follow the approved amended run-in procedures in such instances.

COMPRESSION TESTING OF AIRCRAFT ENGINE CYLINDERS

The purpose of testing cylinder compression is to determine the internal condition of the combustion chamber by ascertaining if any appreciable leakage is occurring.

a. Types of Compression Testers

The two basic types of compression testers currently in use are the direct compression and the differential pressure (Fig. 8-1)

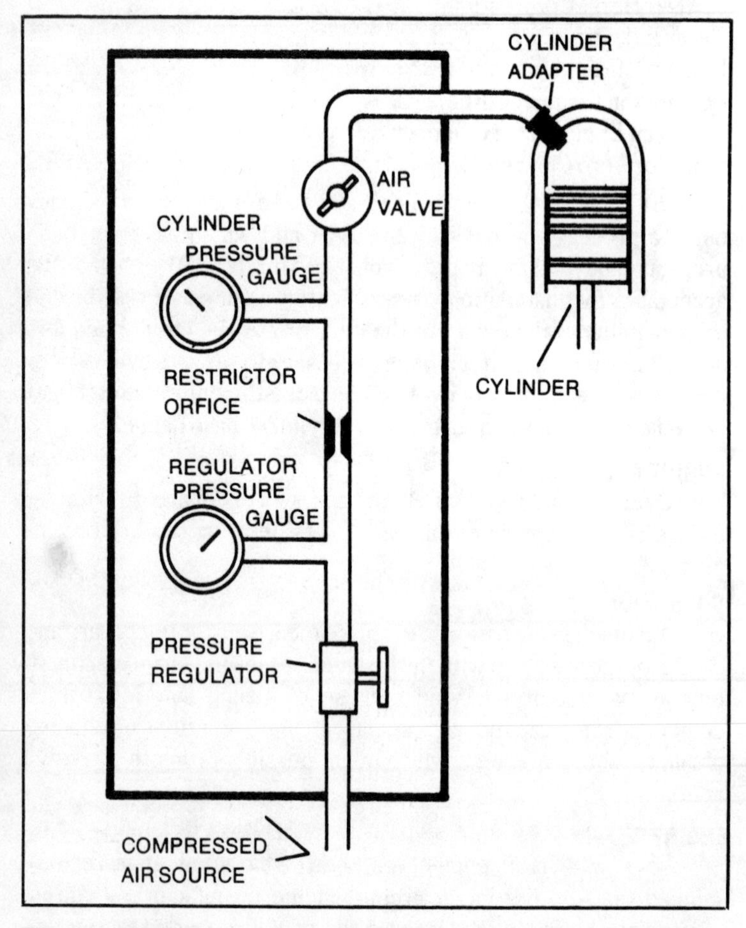

Fig. 8-1. Schematic of a typical differential pressure compression tester (courtesy of FAA).

type testers. The optimum procedure would be to utilize both types of testers when checking the compression of aircraft cylinders. In this respect, it is suggested that the direct compression method be used first and the findings substantiated with the differential pressure method. This provides a cross-reference to validate the readings obtained by each method and tends to assure that the cylinder is defective before it is removed. Before beginning a compression test, consider the following points:

(1) When the spark plugs are removed, identify them to coincide with the cylinder. Close examination of the plugs will reveal the actual operating conditions and aid in diagnosing problems within the cylinder.

(2) The operating and maintenance records of the engine should be reviewed. Records of previous compression tests are of assistance in determining progressive wear conditions and help to establish the necessary maintenance actions.

(3) Before beginning a compression check, precautions should be taken to prevent the accidental starting of the engine.

b. Direct Compression Check

This type of compression test indicates the actual pressures within the cylinder. Although the particular defective component within the cylinder is difficult to determine with this method, the consistency of the readings for all cylinders is an indication of the condition of the engine as a whole. The following are suggested guidelines for performing a direct compression test.

(1) Thoroughly warm up the engine to operating temperatures and perform the test as soon as possible after shutdown.

(2) Remove the most accessible spark plug from each cylinder.

(3) Rotate the engine with the starter to expel any excess oil or loose carbon in the cylinders.

(4) If a complete set of compression testers is available, install one tester in each cylinder. However, if only one tester is being used, check each cylinder in turn.

(5) Rotate the engine at least three complete revolutions using the engine starter and record the compression reading.

NOTE: It is recommended that an external power source be used, if possible, as a low battery will result in a slow engine-turning rate and lower readings. This will noticeably affect the validity of the second engine test on a twin-engine aircraft.

(6) Recheck any cylinder which shows an abnormal reading when compared with the others. Any cylinder having a reading

approximately 15 psi lower than the others should be suspected of being defective.

(7) If a compression tester is suspected of being defective, replace it with one known to be accurate and recheck the compression of the affected cylinders.

c. Differential Pressure Compression Check

The differential pressure tester is designed to check the compression of aircraft engines by measuring the leakage through the cylinders caused by worn or damaged components. The operation of the compression tester is based on the principle that, for any given airflow through a fixed orifice, a constant pressure drop across that orifice will result. The restrictor orifice dimensions in the differential pressure tester should be sized for the particular engine as follows:

1. Engines up to 1000-cubic-inch displacement, .040-orifice diameter, .250-inch long, 60-degree approach angle.

2. Engines in excess of 1000-cubic-inch displacement .060-orifice diameter, .250-inch long, 60 degree approach angle.

As the regulated air pressure is applied to one side of the restrictor orifice with the air valve closed, there will be no leakage on the other side of the orifice and both pressure gauges will read the same. However, when the air valve is opened and leakage through the cylinder increases, the cylinder pressure gauge will record a proportionally lower reading.

(1) Performing the Check. The following procedures are listed to outline the principles involved, and are intended to suppliment the manufacturer's instructions for the particular tester being utilized.

(a) Perform the compression test as soon as possible after the engine is shut down to ensure that the piston rings, cylinder walls and other engine parts are well lubricated.

(b) Remove the most accessible spark plug from each cylinder.

(c) With the air valve closed, apply an external source of clean air (approximately 100 to 120 psi) to the tester.

(d) Install an adapter in the spark plug bushing and connect the compression tester to the cylinder.

(e) Adjust the pressure regulator to obtain a reading of 80 psi on the regulator pressure gauge. At this time, the cylinder pressure gauge should also register 80 psi.

(f) Turn the crankshaft by hand in the direction of rotation until the piston (in the cylinder being checked) is coming up on its compression stroke. Slowly open the air valve and pressurize the cylinder to approximately 20 psi.

CAUTION

Care must be exercised in opening the air valve since sufficient air pressure will have built up in the cylinder to cause it to rotate the crankshaft if the piston is not at TDC.

Continue rotating the engine against this pressure until the piston reaches top dead center (TDC). Reaching TDC is indicated by a flat spot or sudden decrease in force required to turn the crankshaft. If the crankshaft is rotated too far, back up at least one-half revolution and start over again to eliminate the effect of backlash in the valve operating mechanism and to keep piston rings seated on the lower ring lands.

(g) Open the air valve completely. Check the regulated pressure and adjust, if necessary, to 80 psi.

(h) Observe the pressure indication on the cylinder pressure gauge. The difference between this pressure and the pressure shown by the regulator pressure gauge is the amount of leakage through the cylinder. A loss in excess of 25 percent of the input air pressure is cause to suspect the cylinder of being defective. However, recheck the readings after operating the engine for at least 3 minutes to allow for sealing of the rings with oil.

(i) If leakage is still occurring after a recheck, it might be possible to correct a low reading. This is accomplished by placing a fiber drift on the rocker arm directly over the valve stem and tapping the drift several times with a hammer to dislodge any foreign material between the valve face and seat.

NOTE: When correcting a low reading in this manner, rotate the propeller so the piston will not be at TDC. This is necessary to prevent the valve from striking the top of the piston in some engines. Rotate the engine before rechecking compression to reseat the valves in the normal manner.

SPARK PLUGS

The spark plug provides the high voltage electrical spark to ignite the fuel/air mixture in the cylinder. The types of spark plugs used in different engines will vary in regard to heat range, reach, thread size and other characteristics required by the particular installation.

a. Heat range. The heat range of a spark plug is a measure of its ability to transfer heat to the cylinder head. The plug must operate hot enough to burn off the residue deposits which can cause fouling, yet remain cool enough to prevent a preignition condition from occurring. The length of the nose core is the principal factor in establishing the plug's heat range. "Hot" plugs have a long insulator nose and thereby create a long heat transfer path. "Cold" plugs have a relatively short insulator to provide a rapid transfer of heat to the cylinder head (Fig. 8-2).

b. Reach. The spark plug reach (Fig. 8-3) is the threaded portion which is inserted into the spark plug bushing of the cylinder. A plug with the proper reach will insure that the electrode end inside the cylinder is in the best position to achieve ignition. Spark plug seizure or improper combustion within the cylinder will probably occur if a plug with the wrong reach is used.

c. Installation Procedures. When installing spark plugs, observe the following procedure:

(1) Visually inspect the plug for cleanliness and condition of the threads, ceramic, and electrodes. NOTE: Never install a spark plug which has been dropped.

(2) Check the plug for the proper gap setting using a round wire feeler gauge. In the case of used plugs, procedures for cleaning and regapping are usually contained in the various manufacturers' manuals.

(3) Check the plug and cylinder bushing to ascertain that only one gasket is used per spark plug. When a thermocouple-type gasket is used, no other gasket is required.

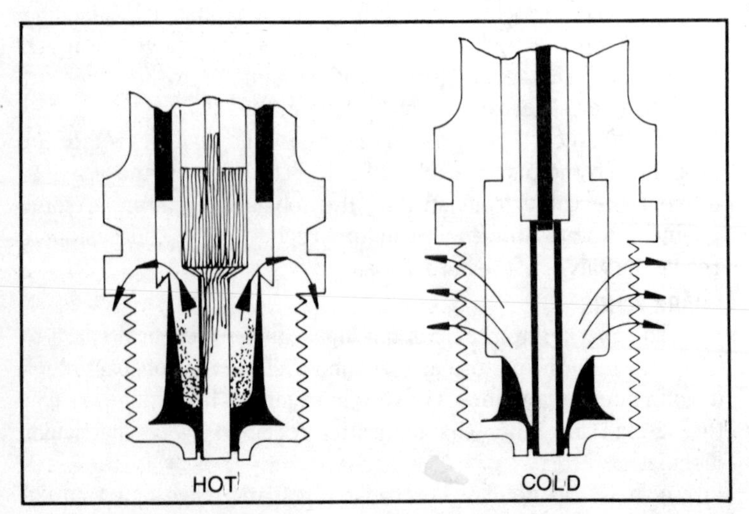

Fig. 8-2. Spark plug heat ranges.

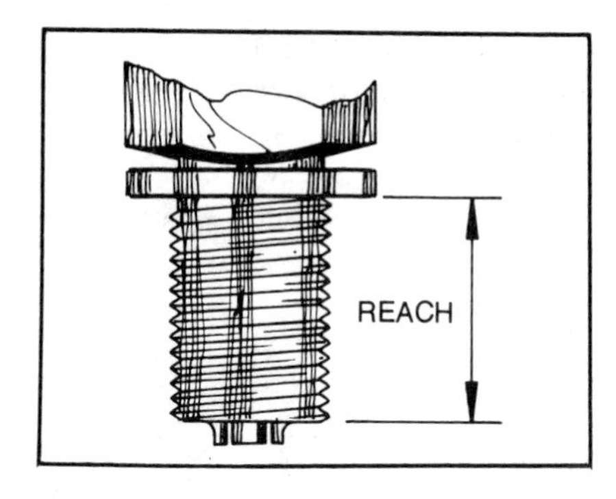

Fig. 8-3. Spark plug reach.

(4) Apply antiseize compound sparingly to the shell threads, but do not allow the compound to contact the electrodes as the material is conductive and will short out the plug. If desired, the use of antiseize compound may be eliminated on engines equipped with stainless steel spark plug bushings or inserts.

(5) Screw the plug into the cylinder head as far as possible by hand. If the plug will not turn easily to within 2 or 3 threads of the gasket, it may be necessary to clean the threads.

NOTE: Cleaning inserts with a tap is not recommended as permanent damage to the insert may result.

(6) Seat the proper socket securely on the spark plug and tighten to the torque limit specified by the engine manufacturer before proceeding to the next plug.

CAUTION

A loose spark plug will not transfer heat properly and, during engine operation, could overheat to the point where the nose ceramic will become a "hot spot" and cause preignition. However, avoid overtightening. Damage to the plug and bushing could result.

(7) Connect the ignition lead after wiping clean with methylethylketone (MEK), acetone, or similar material. Insert the terminal assembly into the spark plug in a straight line. (Care should be taken as improper techniques can damage the terminal sleeves.) Screw the connector nut into place until finger tight; then tighten an additional one-quarter turn while holding the elbow in the proper position.

(8) Perform an engine runup after installing a new set of spark plugs. When the engine has reached normal operating temperatures, check the magnetos and spark plugs in accordance with the manufacturer's instructions.

d. Operational Problems. Whenever problems develop during engine operation which appear to be caused by the ignition system, it is recommended that the spark plugs and ignition harnesses be checked first before working on the magnetos. The following are the most common spark plug malfunctions and are relatively easy to identify.

(1) Fouling

(a) Carbon fouling. This is identifiable by the dull black, sooty deposits on the electrode end of the plug (Fig. 8-4). Although the primary causes are excessive ground idling and rich idle mixtures, plugs with a cold heat range might also be a contributing factor.

(b) Lead fouling. This is characterized by hard, dark, cinderlike globules which gradually fill (Fig. 8-5) up the electrode cavity and short out the plug. The primary cause for this condition is poor fuel vaporization combined with a high tetraethyl-lead content fuel. Plugs with a cold heat range might also contribute to this condition.

(c) Oil fouling. This is identifiable by a wet, black carbon deposit over the entire firing end of the plug. Fig. 8-6. This condition is fairly common on the lower plugs in horizontally opposed engines and on both plugs in the lower cylinders of radial engines. Oil fouling is normally caused by oil drainage past the piston rings after shutdown. However, when both spark plugs removed from the same cylinder are badly fouled with oil and carbon, some form of engine damage should be suspected and the cylinder should be more closely inspected. Mild forms of oil fouling can usually be cleared up by slowly increasing power while running the engine until the deposits are burned off and the misfiring stops.

(2) Copper Runout

Some erosion or "cupping" of the exposed center electrode copper core is normal and will gradually decrease with the service life of the plug until it practically ceases. This condition (Fig. 8-7) is not a cause for rejection until the erosion has progressed to a point more than 3/32 of an inch below the tip of the center electrode.

The high temperatures and pressures associated with preignition can cause the condition shown in Fig. 8-8. In this instance, the copper center electrode melted and flowed out, bridged the electrodes, and caused a shorted condition.

Fig. 8-4. Typical carbon fouled spark plug.

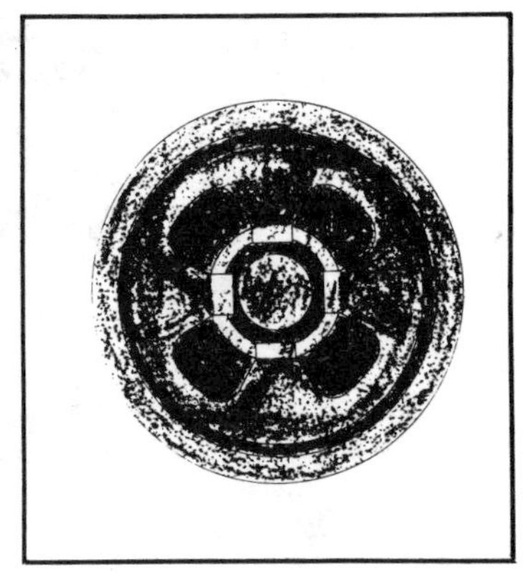

Fig. 8-5. Typical lead fouled spark plug.

(3) Bridged Electrodes

Occasionally, free combustion chamber particles will settle on the electrodes of a spark plug and gradually bridge the electrode gap. The result is a shorted plug. Small particles can be dislodged by slowly cycling the engine as described for the oil-fouled condition. However, the only remedy for more advanced cases is removal and replacement of the spark plug. This condition is shown in figure 8-9.

(4) Metal Deposits

Whenever metal spray is found on the electrodes of a spark plug, it is an indication that a failure of some part of the engine is in

Fig. 8-6. Typical oil fouled spark plug.

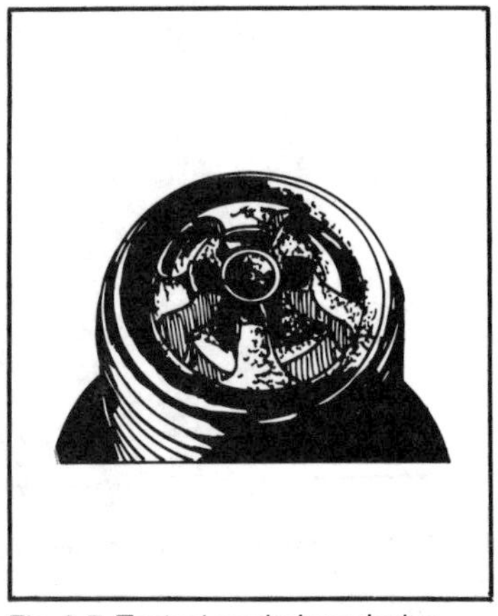

Fig. 8-7. Typical eroded spark plug.

Fig. 8-8. Typical spark plug with copper runout.

progress. The location of the cylinder in which the spray is found is important in diagnosing the problem. Various types of failures will cause the metal spray to appear differently. For example, if the metal spray is located evenly in every cylinder, the problem will be in the induction system (such as an impeller failure). If the metal spray is only found on the spark plugs in one cylinder, the problem is isolated to that cylinder and will generally be a piston failure. In view of the secondary damage which occurs whenever an engine part fails, any preliminary indication such as metal spray should be thoroughly investigated to establish the cause and correct it.

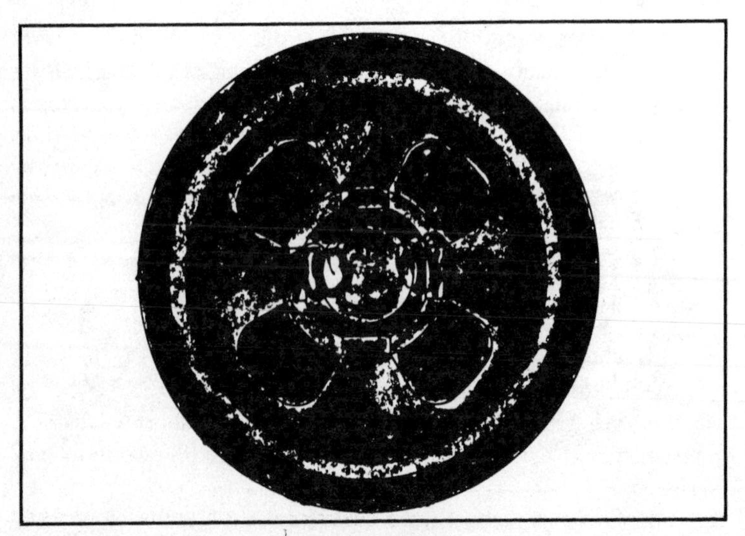

Fig. 8-9. Typical spark plug with bridged electrodes.

(5) Flashover

It is important that spark plug terminal contact springs and moisture seals be checked regularly for condition and cleanness to prevent "flashover" in the connector well. Foreign matter or moisture in the terminal connector well can reduce the insulation value of the connector to the point where ignition system voltages at higher power settings could flash over the connector well surface to ground and cause the plug to misfire. If moisture is the cause, hard starting can also result. The cutaway spark plug shown in Fig. 8-10 illustrates this malfunction. Any spark plug found with a dirty connector could have this condition and should be reconditioned before reuse.

Pre-reconditioning Inspection

All spark plugs should be inspected visually prior to reconditioning to eliminate any plug with obvious defects. A partial checklist of common defects includes:

(1) Ceramic chipped or cracked either at the nose core or in the connector well.

(2) Damaged or badly worn electrodes.

(3) Badly nicked, damaged or corroded threads on shell or shielding barrel.

(4) Shielding barrel dented, bent or cracked.

(5) Connector seat at the top of the shielding barrel badly nicked or corroded.

IGNITION HARNESSES

Aircraft quality ignition harness is usually made of either medium or high temperature wire. The type used will depend upon the manufacturing specification for the particular engine. In addition to the applicable manufacturer's maintenance and repair procedures, the following is a quick-reference checklist for isolating some of the malfunctions inherent to ignition harnesses.

a. Carefully inspect the lead conduit or shielding. A few broken strands will not affect serviceability. If the insulation in general looks worn, replace the lead.

b. When replacing a lead, if the dressing procedure is not accomplished properly, strands of shielding could be forced through the conductor insulation. If this occurs, a short will exist in the conductor. Therefore, it is essential that this task be performed properly.

c. The high temperature coating used on some lightweight harnesses is provided for vibration abrasion resistance and moisture protection. Slight flaking or peeling of this coating is not serious and a harness assembly need not be removed from service because of this condition.

d. Check the spark plug contact springs for breaks, corrosion or deformation. If possible, check the lead continuity from the distributor block to the contact spring.

e. Check the insulators at the spark plug end of the lead for cracks, breaks or evidence of old age. Make sure they are clean.

f. Check to see that the leads are positioned as far away from the exhaust manifold as possible and are supported to prevent any whipping action.

g. When lightweight harnesses are used and the conduit enters the spark plug at a severe angle, use clamps as shown in Fig. 8-11 to prevent overstressing the lead.

MAGNETO INSPECTION

Whenever ignition problems develop and it is determined that the magneto is the cause of the difficulty, the following are a few simple inspection procedures which might locate the malfunction quickly. However, conduct any internal inspection or repair of a magneto in accordance with the manufacturer's maintenance and overhaul manuals.

a. Inspect the distributor block contact springs. Replace them if they are broken or corroded.

b. Inspect the felt oil washer if applicable. It should be saturated with oil. If it is dry, check for a worn bushing.

c. Inspect the distributor block for cracks or a burned area. The wax coating on the block should not be removed. Do not use any solvents for cleaning.

d. Look for excess oil in the breaker compartment. If oil is present, it might mean a bad oil seal or bushing at the drive end.

Fig. 8-10. Spark plug well flushover.

This condition could require complete overhaul. Too much oil can foul and cause excessive burning of the contact points.

e. Look for frayed insulation of the leads in the breaker compartment of the magneto. See that all terminals are secure. Be sure that wires are properly positioned.

f. Inspect the capacitor visually for general condition and check the mounting bracket for cracks or looseness. If possible, check the capacitor for leakage, capacity and series resistance.

g. Examine the points for excessive wear or burning. Discard points which have deep pits or excessively burned areas. Desired contact surfaces have a dull gray, sandblasted (almost rough) or frosted appearance over the area where electrical contact is made. Minor irregularities or roughness of point surfaces are not harmful. Neither are small pits or mounds if they are not too pronounced. If there is a possibility of the pit becoming deep enough to penetrate the pad, reject the contact assembly.

Generally, no attempt should be made to dress or stone contact point assemblies. However, if provided, procedures and limits contained in the manufacturer's manuals should be followed.

FUEL SYSTEMS

When fuel system lines are to be replaced or repaired, consider the following fundamentals in addition to the applicable airworthiness requirements.

a. Compatibility of Fittings. All fittings are to be comparible with their mating parts. Although various types of fittings appear to be interchangeable, in many cases they have different thread pitch or minor design differences which prevent proper mating and could cause the joint to leak or fail.

b. Routing. Make sure that the line does not chafe against control cables, airframe structure, etc., or come in contact with electrical wiring or conduit. Where physical separation of the fuel lines form electrical wiring or conduit is impracticable, locate the fuel line below the wiring and clamp it securely to the airframe structure. In no case should wiring be supported by the fuel line.

c. Alignment. Locate bends accurately so that the tubing is aligned with all support clamps and end fittings and is not drawn, pulled or otherwise forced into place by them. Never install a straight length of tubing between two rigidly mounted fittings. Always incorporate at least one bend between such fittings to absorb strain caused by vibration and temperature changes.

d. Bonding. Bond metallic fuel lines at each point where they are clamped to the structure. Integrally bonded and cushioned line support clamps are preferred to other clamping and bonding methods.

e. Support of Line Units. To prevent possible failure, all fittings heavy enough to cause the line to sag should be supported by means other than the tubing.

f. Support Clamps. Place support clamps or brackets for metallic lines as follows:

Tube O.D.	Approximate distance between supports
⅛"-3/16"	9"
¼"-5/16"	12"
⅜"-½"	16"
⅝"-¾"	22"
1"-1¼"	30"
1½"-2"	40"

Locate clamps or brackets as close to bends as possible to reduce overhang (Fig. 8-12).

FUEL TANKS AND CELLS

Welded or riveted fuel tanks that are made of commercially pure aluminum, 3003, 5052, or similar alloys, can be repaired by welding. Tanks made from heat-treatable aluminum alloys are generally assembled by riveting. In case it is necessary to rivet a new piece in a tank, use the same material as used in the tank

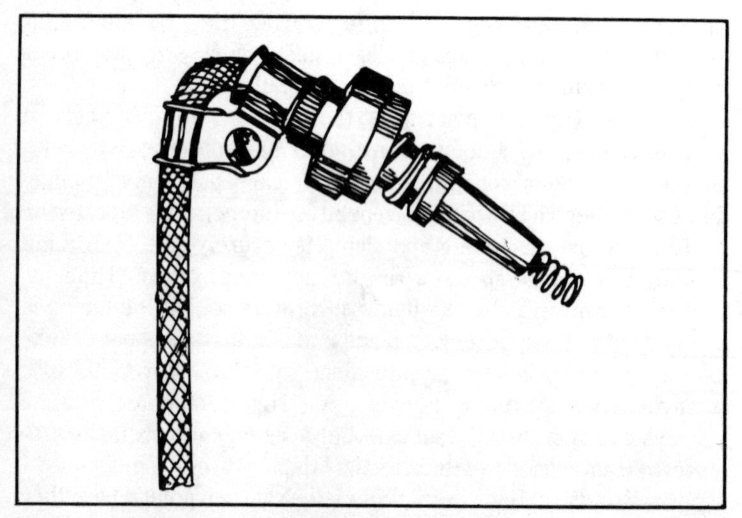

Fig. 8-11. Typical method of clamping leads.

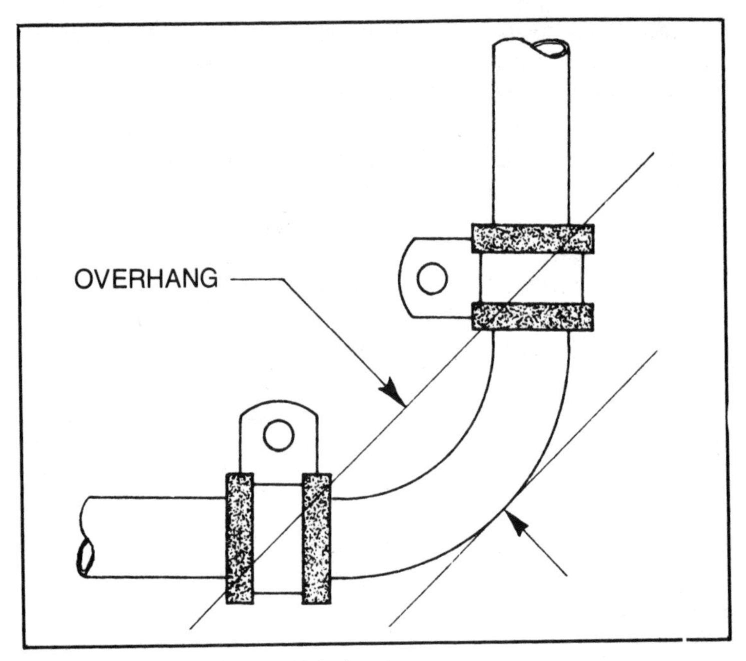

Fig. 8-12. Location of clamps at tube bends.

undergoing repair and seal the seams with a compound that is insoluble in gasoline. Special sealing compounds are available and should be used in the repair of tanks. Inspect fuel tanks and cells for general condition, security of attachment and evidence of leakage. Examine fuel tank or cell vent line, fuel line and sump drain attachment fittings closely.

CAUTION

Purge defueled tanks of explosive fuel/air mixtures in accordance with the manufacturers' service instructions. In the absence of such instructions, utilize an inert gas such as CO_2 as a purgative to assure total deletion of fuel/air mixtures.

a. Integral Tanks. Examine the interior surfaces and seams for sealant deterioration and corrosion (especially in the sump area). Follow the manufacturer's instructions for repair and cleaning procedures.

b. Internal Metal Tanks. Check the exterior for corrosion and chafing. Dents or other distortion, such as a partially collapsed tank caused by an obstructed fuel tank vent, can adversely affect fuel quantity gauge accuracy and tank capacity. Check the interior surfaces for corrosion. Pay particular attention to the sump area (especially those which are made of cast material).

c. Removal of Flux After Welding. It is especially important, after repair by welding, to completely remove all flux in order to avoid possible corrosion. Promptly upon completion of welding, wash the inside and outside of the tank with liberal qnantities of hot water and then drain. Next, immerse the tank in either a 5-percent nitric or 5-percent sulfuric acid solution. If the tank cannot be immersed, fill the tank with either solution and wash the outside with the same solution. Permit the acid to remain in contact with the weld about one hour and then rinse thoroughly with clean water. Test the efficiency of the cleaning operation by applying some acidified 5-percent silver nitrate solution to a small quantity of the rinse water used last to wash the tank. If a heavy white precipitate is formed, the cleaning is insufficient and the washing should be repeated.

d. Flexible Fuel Cells. Inspect the interior for checking, cracking, porosity or other signs of deterioration. Make sure the cell retaining fasteners are properly positioned. If repair or further inspection is required, follow the manufacturer's instructions for cell removal, repair and installation. Do not allow flexible fuel cells to dry out. Preserve them in accordance with the manufacturer's instructions.

FUEL TANK CAPS, VENTS, AND OVERFLOW LINES

Inspect the fuel tank caps to determine that they are the correct type and size for the installation.

a. Vented Caps. If these are substituted for unvented caps, they could cause loss of fuel or fuel starvation. Similarly, an improperly installed cap that has a special venting arrangement can also cause malfunctions.

b. Unvented Caps. If these are substituted for vented caps, they will cause fuel starvation and possible collapse of the fuel tank or cell. Malfunctioning of this type occurs when the pressure within the tank decreases as the fuel is withdrawn. Eventually, a point is reached where the fuel will no longer flow and the outside atmospheric pressure collapses the tank. As a result, the effects will occur sooner with a full fuel tank than with one partially filled.

c. Check Tank Vents And Overflow Lines. Inspect for condition, obstructions, correct installation and proper operation of any check valves and ice protection units. Pay particular attention to the location of the tank vents when such information is provided in the manufacturer's service instructions. Inspect for cracked or deteriorated filler opening recess drains which could allow spilled fuel to accumulate within the wing or fuselage. One method of inspection is to plug the fuel line at the outlet and observe fuel placed in the filler opening recess. If drainage takes place, investigate condition of the line and purge any excess fuel from the wing.

d. Filler Marks. Assure that filler opening markings are stated according to the applicable airworthiness requirements and are complete and legible.

FUEL CROSSFEED, FIREWALL SHUTOFF AND TANK SELECTOR VALVES

Inspect these valves for leakage and proper operation as follows:

a. Internal Leakage. This can be checked by placing the appropriate valve in the "off" position, draining the fuel strainer bowl and observing if fuel continues to flow into it. Check all valves located downstream of boost pumps with the pump(s) operating. Do not operate the pump(s) longer than necessary.

b. External Leakage. This can be a severe fire hazard and this is especially true if the unit is located under the cabin floor or within a similarly confined area. Correct the cause of any fuel stains associated with fuel leakage.

c. Selector Handles. Check the operation of each handle or control to see that it indicates the actual position of the selector valve. Assure that stops and detents have positive action and feel. Worn or missing detents and stops can cause unreliable positioning of the fuel selector valve.

d. Worn Linkage. Inaccurate positioning of fuel selector valves can also be caused by worn mechanical linkage between the selector handle and the valve unit. An improper fuel valve position setting can seriously reduce engine power by restricting the available fuel flow. Check universal joints, pins, gears, splines, cams, levers, etc., for wear and excessive clearance which prevent the valve from positioning accurately or from obtaining fully "off" and "on" positions.

e. Placards. Assure that required placards are complete and legible. Replace those that are missing or cannot be read easily.

FUEL PUMPS

Inspect, repair and overhaul boost pumps, emergency pumps, auxiliary pumps and engine-driven pumps in accordance with the appropriate manufacturer's instructions.

FUEL FILTERS, STRAINERS AND DRAINS

Check each strainer and filter element for contamination. Determine and correct the source of any contaminants found. Replace throwaway filter elements with the recommended type. Examine fuel strainer bowls to see that they are properly installed according to direction of fuel flow. Check the operation of all drain devices to see that they operate properly and have positive shutoff action.

INDICATOR SYSTEMS

Inspect, service and adjust the fuel indicator systems according to the manufacturer's instructions. Determine that the required placards and instrument markings are complete and legible.

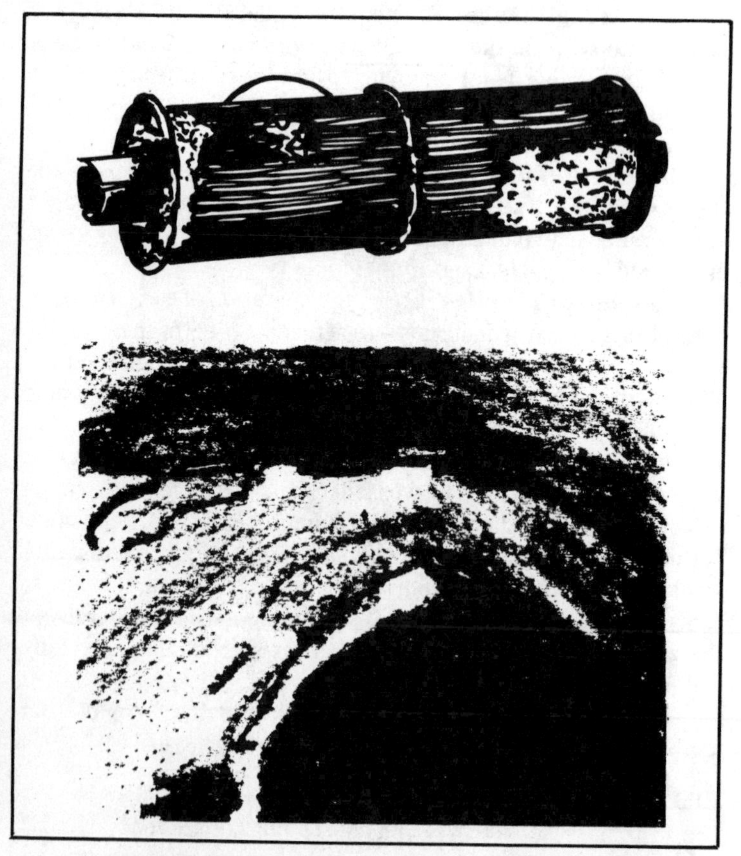

Fig. 8-13. Typical muffler wall fatigue failure at exhaust outlet. (A. Complete muffler assembly with heat shroud removed. B. Detail view of failure).

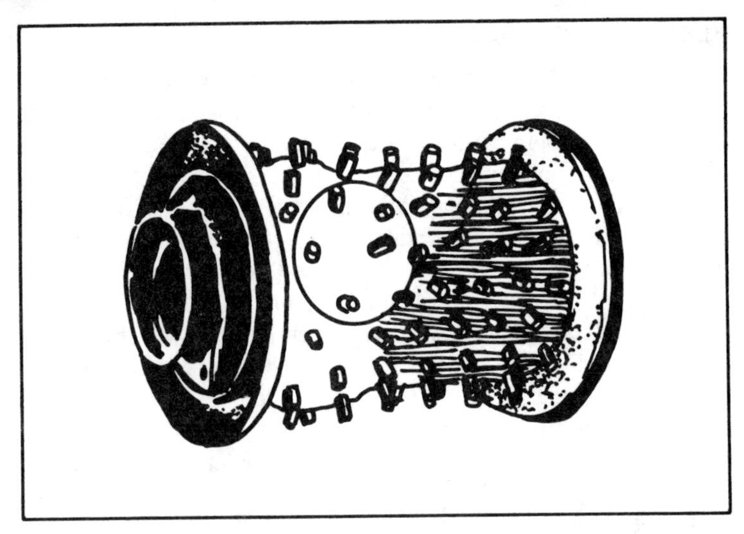

Fig. 8-14A. Typical muffler wall failure. Complete muffler assembly with heat shroud removed.

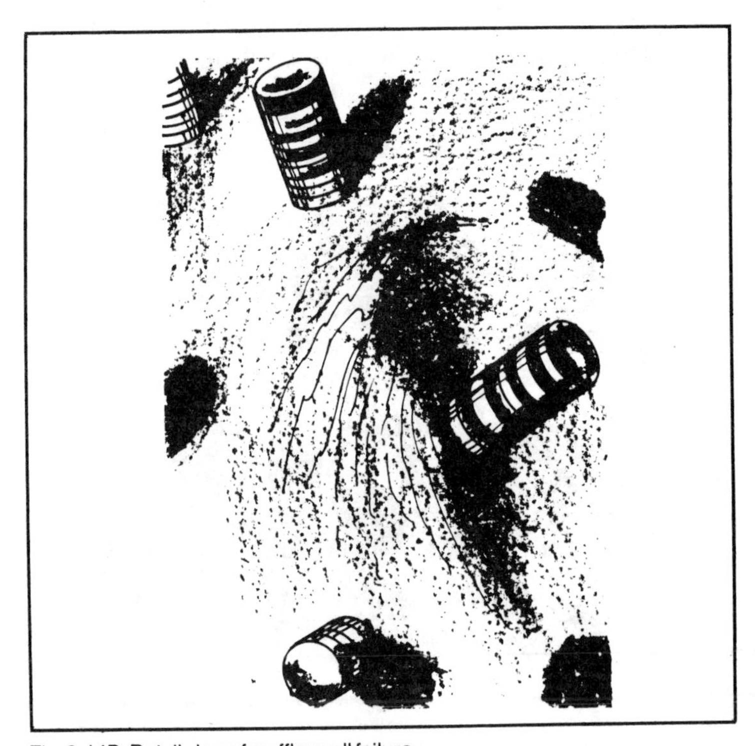

Fig. 8-14B. Detail view of muffler wall failure.

TURBINE FUEL SYSTEMS

The use of turbine fuels in aircraft has resulted in two problem areas not normally associated with aviation gasolines: entrained water (microscopic particles of free water suspended in the fuel) and microbial contaminants.

a. Entrained Water. This will remain suspended in aviation turbine fuels for a considerable length of time. Unless suitable measures are taken, the fine filters used in turbine fuel systems will clog with ice crystals when the temperature of the fuel drops below the freezing temperature of the entrained water. Some fuel systems employ heated fuel filters or fuel heaters to eliminate this problem. Others rely upon anti-icing fuel additives.

b. Microbial Contamination. This is a relatively recent problem associated with the operation and maintenance of turbine engine fuel systems. The effects of these microorganisms are far reaching. They can cause powerplant failure due to clogging of filters, lines, fuel controls, ets., and the corrosive acids which they produce can lead to structural failure of integral tanks. Microbial contamination is generally associated with fuel containing free water introduced by condensation or other extraneous sources.

c. Maintenance. Maintain turbine engine fuel systems and use anti-icing and antibacterial additives in accordance with the manufacturer's recommendations.

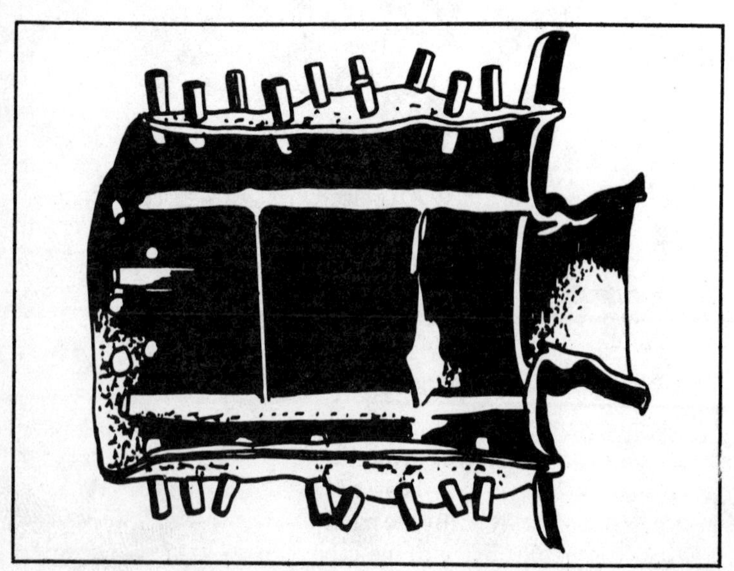

Fig. 8-14C. Cross section of a failed muffler.

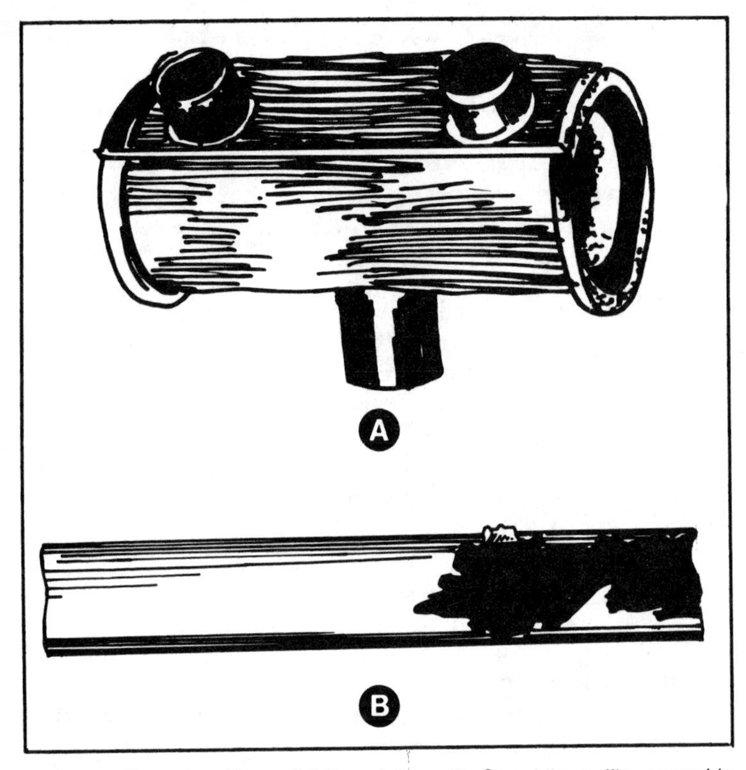

Fig. 8-15. Typical muffler wall fatigue failure. (A. Complete muffler assembly with heat shroud partially removed. B. Detail view of failure).

EXHAUST SYSTEMS

Any exhaust system failure should be regarded as a severe hazard. Depending upon the location and type of failure, it can result in carbon monoxide (CO) poisoning of crew and passengers, partial or complete engine power loss or fire. Exhaust system failures generally reach a maximum rate of occurrence at 100 to 200 hours' operating time and over 50 percent of the failures occur within 400 hours.

Muffler/Heat Exchanger Failures

Approximately one-half of all exhaust system failures are traced to cracks or ruptures in the heat exchanger surfaces used for cabin and carburetor air heat sources. Failures in the heat exchanger's surface (usually the muffler's outer wall) allow exhaust gases to escape directly into the cabin heat system. The failures are, for the most part, attributed to thermal and vibration fatigue cracking in areas of stress concentration; e.g., tailpipe and stack inlet attachment areas (Figs. 8-13 through 8-16).

Failures of the spot welds which attach heat transfer pins, as shown in Fig. 8-14(A), can result in exhaust gas leakage. In addition to the carbon monoxide hazard, failure of heat exchanger surfaces can permit exhaust gases to be drawn into the engine induction system and cause engine overheating and power loss.

Manifold/Stack Failures

Exhaust manifold and stack failures are also usually fatigue-type failures which occur at welded or clamped joints; e.g., stack-to-flange, stack-to-manifold, muffler connections or crossover pipe connections. Although these failures are primarily a fire hazard, they also present a carbon monoxide problem. Exhaust gases can enter the cabin via defective or inadequate seals at firewall openings, wing strut fittings, doors and wing root openings. Manifold/stack failures, which account for approximately 20 percent of all exhaust system failures, reach a maximum rate of occurence at about 100 hours' operating time. Over 50 percent of the failures occur within 300 hours.

Internal Muffler Failures

Internal failures (baffles, diffusers, etc.) can cause partial or complete engine power loss by restricting the flow of the exhaust gases (Figs. 8-17 through 8-20). As opposed to other failures, erosion and carbonizing caused by the extreme thermal conditions are the primary causes of internal failures. Engine backfiring and combustion of unburned fuel within the exhaust system are

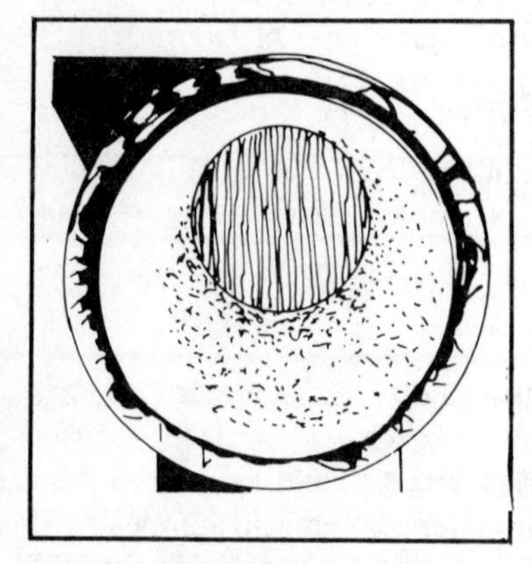

Fig. 8-16. Typical fatigue failure of muffler end plate at stack inlet.

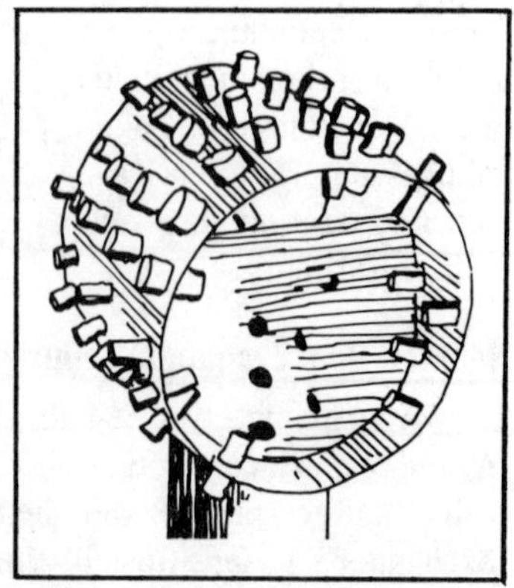

Fig. 8-17. Section of muffler showing typical internal baffle failure.

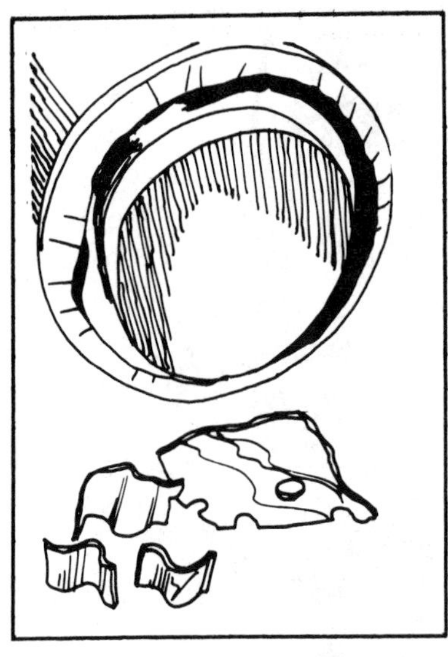

Fig. 8-18. Loose pieces of failed internal baffle.

probable contributing factors. In addition, local hot spot areas caused by uneven exhaust gas flow results in burning, bulging and rupture of the outer muffler wall (Fig 8-14). As might be expected, the time required for these failures to develop is longer than that for fatigue failures. Internal muffler failures account for nearly 20 percent of the total number of exhaust system failures. The highest rate of internal failures occur between 500 and 750 hours of operating time. Engine power loss and excessive backpressure caused by exhaust outlet blockage can be averted by the installation of an exhaust outlet guard as shown in Figs. 8-21 and 8-22.

The outlet guard can be fabricated from 3/16″ stainless steel welding rod. Form the rod into two "U" shaped segments, approximately 3 inches long and weld into the exhaust tail pipe as shown in Fig. 8-21 so that the guard will extend 2 inches inside the exhaust muffler outlet port. Installation of an exhaust outlet guard does not negate the importance of thorough inspection of the internal parts of the muffler or the necessity of replacing defective mufflers.

Inspection

Inspect exhaust systems frequently to ascertain complete system integrity.

CAUTION

Marking of exhaust system parts—NEVER USE LEAD PENCILS, GREASE PENCILS, ETC., TO MARK EXHAUST SYSTEM PARTS. Carbon deposited by those tools will

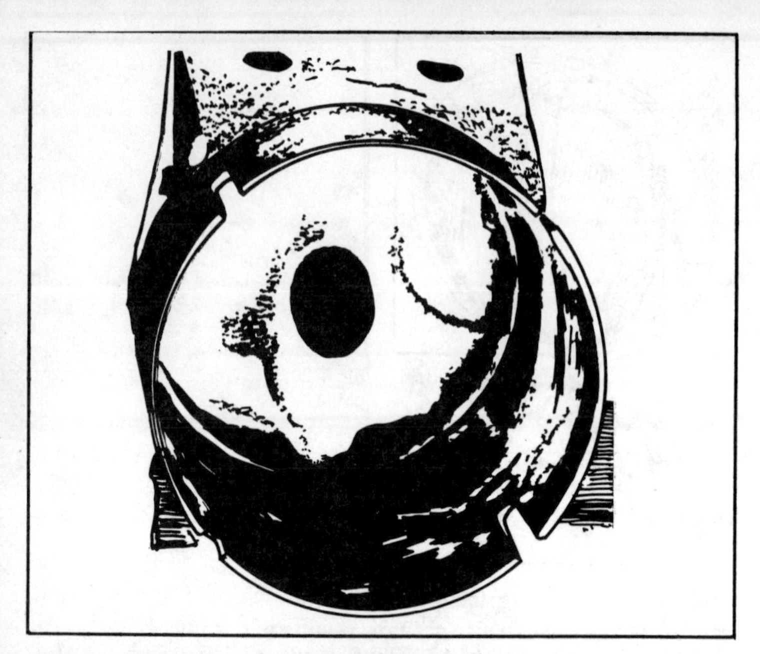

Fig. 8-19. Failed internal baffle partially obstructing muffler outlet.

Fig. 8-20. Failed internal baffle completely obstructing muffler outlet.

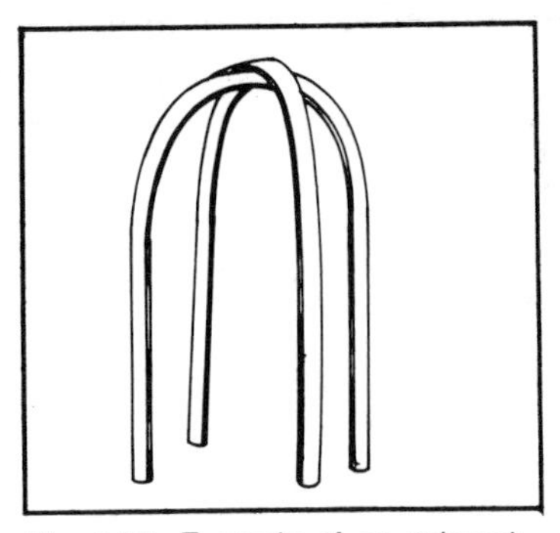

Fig. 8-21. Example of an exhaust outlet guard.

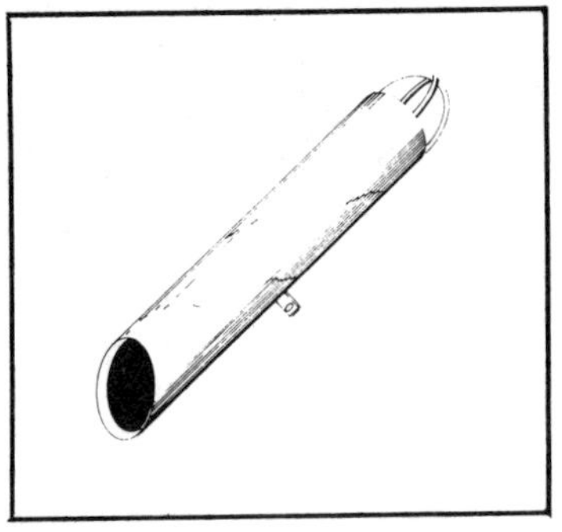

Fig. 8-22. Example of an exhaust outlet guard installed.

cause cracks from heat concentration and carbonization of the metal. If you must mark on exhaust system parts, USE CHALK, PRUSSIAN BLUE OR INDIA INK that is carbon free.

Prior to any cleaning operation, remove cowling as required to expose the complete exhaust system. Examine cowling and nacelle areas adjacent to exhaust system components for telltale signs of exhaust gas soot indicating possible leakage points. Check to make sure that no portion of the exhaust system is being chafed by cowling, engine control cables or other components.

Perform the necessary cleaning operations and inspect all external surfaces of the exhaust system for cracks, dents and missing parts. Pay particular attention to welds, clamps, supports,

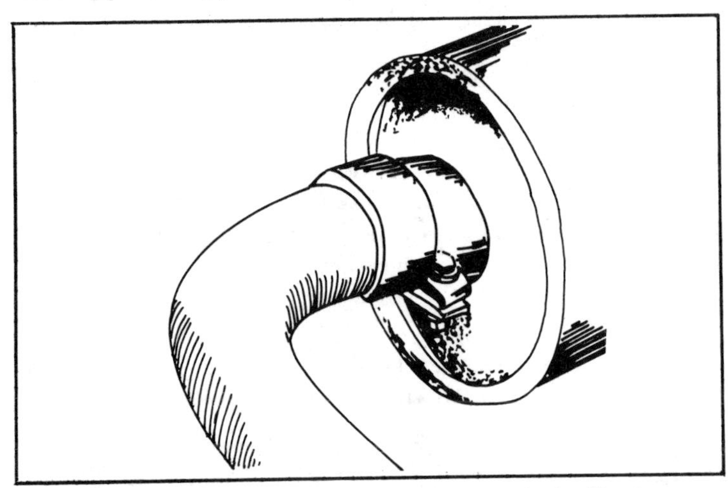

Fig. 8-23. The effect of an improperly positioned exhaust pipe/muffler clamp.

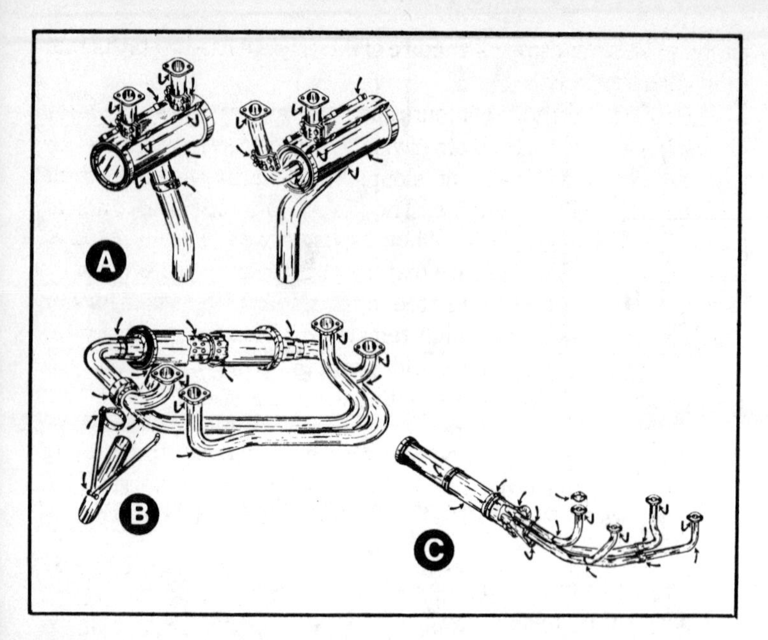

Fig. 8-24. Primary inspection areas (A. separate system; B. crossover type system; C. exhaust/augmentor system).

and support attachment lugs, bracing, slip joints, stack flanges, gaskets, flexible couplings, etc. (See Figs. 8-23 and 8-24.) Examine the heel of each bend, areas adjacent to welds, any dented areas and low spots in the system for thinning and pitting due to internal erosion by combustion products or accumulated moisture. An ice pick or similar pointed instrument is useful in probing suspected areas. Disassemble the system as necessary to inspect internal baffles or diffusers.

Should a component be inaccessible for a thorough visual inspection or hidden by non-removable parts, remove it and check for possible leaks by plugging its openings, applying approximately 2 psi internal pressure and submerging it in water. Any leaks will cause bubbles that can be readily detected. Dry thoroughly before reinstallation.

Repairs

It is generally recommended that exhaust stacks, mufflers, tailpipes, etc., be replaced with new or reconditioned components rather than repaired. Welded repairs to exhaust systems are complicated by the difficulty of accurately identifying the base metal so that the proper repair materials can be selected. Changes in composition and grain structure of the original base metal further complicate the repair.

Retain the original contours and make sure that the completed repair has not warped or otherwise affected the alignment of the exhaust system. Repairs or sloppy weld beads which protrude internally are not acceptable. They cause local hot spots and can restrict exhaust gas flow. When repairing or replacing exhaust system components, be sure that the proper hardware and clamps are used. Do not substitute steel or low temperature self-locking nuts for brass or special high temperature locknuts used by the manufacturer. Never reuse old gaskets. When disassembly is necessary, replace gaskets with new ones of the same type provided by the manufacturer.

Turbosupercharger

When a turbosupercharger is included, the exhaust system operates under greatly increased pressure and temperature conditions. Extra precautions should be taken in the exhaust system's care and maintenance. During high altitude operation, the exhaust system pressure is maintained at or near sea level values. Due to the pressure differential, any leaks in the system will allow the exhaust gases to escape with torchlike intensity that can severely damage adjacent structures. A common cause of turbosupercharger malfunction is coke deposits (carbon buildup) in the waste gate unit causing erratic system operation. Excessive deposit buildups may cause the waste gate valve to stick in the closed position and cause an overboost condition.

Coke deposit buildup in the turbosupercharger itself will cause a gradual loss of power in flight and low deck pressure reading prior to takeoff. Experience has shown that periodic decoking, or removal of carbon deposits, is necessary to maintain peak efficiency. Clean, repair, overhaul and adjust turbosupercharger system components and controls in accordance with the applicable manufacturer's instructions.

Augmentor Systems

Inspect augmentor tubes periodically for proper alignment, security of attachment and general overall condition. Regardless of whether or not the augmentor tubes contain heat exchanger surfaces, they should be inspected for cracks along with the remainder of the exhaust system. Cracks can present a fire or carbon monoxide hazard by allowing exhaust gases to enter nacelle, wing or cabin areas.

LANCASHIRE LIBRARY

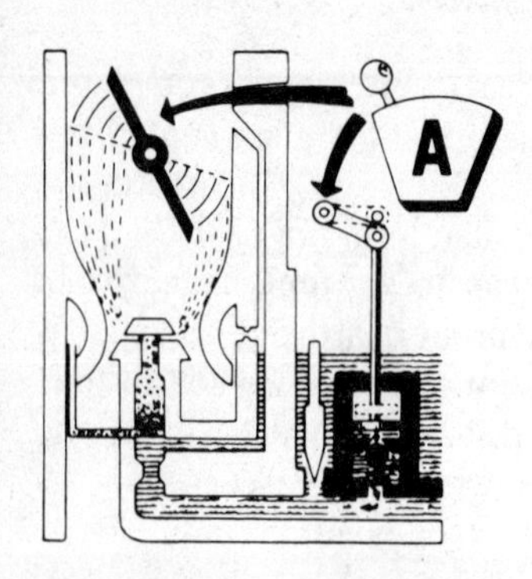

General Aviation Manufacturers

The General Aviation Manufacturers Association (GAMA) member companies account for more than 95 percent of the general aviation production in the United States. They manufacture engines, avionics, airframes and major components of general aviation aircraft. The address for the GAMA is Suite 1215, 1025 Connecticut Ave., N.W., Washington, D.C. 20036, telephone (202) 296-8848. Following is a list of GAMA member companies.

Aero Products Research, Inc.
11210 Hindry Ave.,
Los Angeles, CA 90045
(213) 776-1576

Analog Training Computers, Inc.
189 Monmouth Parkway,
W. Long Branch, NJ 07764
(201) 970-9200

Avco Corporation
Avco Lycoming Division
652 Oliver Street,
Williamsport, PA 17701
(717) 323-6181

Beech Aircraft Corporation
9709 E. Central Ave.,
Wichita, KS 67201
(316) 689-7692

Bendix Corporation
Bendix Center,
Southfield, MI 48075
(313) 352-5000

Cessna Aircraft Company
P. O. Box 1521,
Wichita, KS 67201
(316) 685-9111

Champion Spark Plug Company
P. O. Box 910,
Toledo, OH 43601
(419) 536-3711

Collins Radio Company
Collins Road, N.E.,
Cedar Rapids, IA 52406
(319) 395-1000

Edo-Aire Division
Edo Corporation
216 Passaic Avenue,
Fairfield, NJ 07006
(201) 228-1880

Flite-Tronics Co., Inc.
3314 Burton Avenue,
Burbank, CA 91504
(213) 849-1552

Garret Corporation
9851 Sepulveda Blvd.,
Los Angeles, CA 90009
(213)776-1010

Gates Learjet Corporation
P. O. Box 7707,
Wichita, KS 67277
(316) 946-2345

Grumman Aerospace Corporation
Bethpage, Long Island, NY 11714
(516) 575-0574

King Radio Corporation
400 North Rogers Road
Olathe, KS 66061
(913) 782-0405

Narco Scientific Industries
Commerce Drive,
Fort Washington, PA 19034
(215) 643-2900

North American Rockwell Corp.
General Aviation Divisions
5001 N. Rockwell Ave.,
Bethany, OK 73008
(405) 789-5000

Northrop Airport Development
801 Follin Lane,
Vienna, VA 22180
(703) 938-2070

Oberdorfer Foundries, Inc.
P. O. Box 1125,
Syracuse, NY 13201
(315) 463-3361

Pacific Scientific Company
1346 S. State College Blvd.,
Anaheim, CA 92803
(714) 774-5217

Piper Aircraft Corporation
Lock Haven, PA 17745
(717) 748-6711

PPG Industries, Inc.
777 State National Bank Bldg.,
Huntsville, AL 35801
(205) 539-8121

RCA Corporation
Aviation Equipment Department
8500 Balboa Blvd.,
Van Nuys, CA 91409
(213) 894-8111

Singer Company
Aerospace & Marine
Systems Group
30 Rockefeller Plaza,
New York, NY 10020
(212) 581-4800

Sperry Rand Corporation
Flight Systems Division
P.O. Box 2529,
Phoenix, AZ 85002
(602) 942-2311

Teledyne Continental Motors
30500 Van Dyke Avenue,
Warren, MI 48093
(313) 751-7000

Ub-Flight Devices Corporation
6601 Huntley Road,
Columbus, OH 43229
(614) 846-4300

United Aircraft Corporation
United Aircraft of Canada, Ltd.
P.O. Box 10,
Longueuil, P.Q., Canada
(514) 677-9411

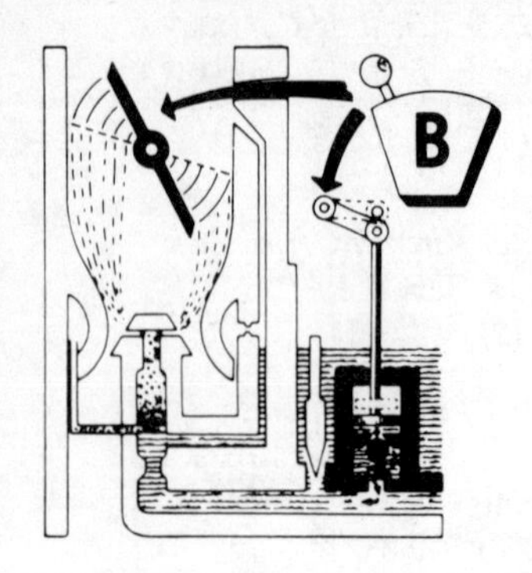

Aviation Organizations

Following are some of the largest and most representative of the hundreds of aviation organizations throughout the United States.

Aerobatic Club of America
% EAA
P.O. Box 229,
Hales Corners, WI 53130
(414) 425-4860 or 4871

Aerobatic Club of America
P.O. Box 401,
Roanoke, TX 76262

Aerospace Education Foundation
1750 Pennsylvania Ave., N.W.,
Washington, D.C. 20006
(202) 298-9123

Aerospace Industries Association of America (AIA
1725 DeSales Street, N.W.,
Washington, D.C. 20036
(202) 347-2315

Aircraft Electronics Association
6310 Gen. Twinning Avenue,
Sarasota, FL 33580
(813) 355-7625

Aircraft Owners & Pilots Assoc.
7315 Wisconsin Avenue,
Bethesda, MD
Mail to: P. O. Box 5800,
Washington, D.C. 20014
(202) 654-0500

Air Force Association
1750 Pennsylvania Ave., N.W.,
Washington, D.C. 20006
(202) 298-9123

Airport Operators Council Intl.
1700 K Street, N.W.,
Washington, D.C. 20006
(202) 296-3270

Air Traffic Control Assoc.
Suite 409, ARBA Bldg.,
525 School Street, S.W.,
Washington, D.C. 20024
(202) 347-5100

Air Transport Association of America (ATA)
1000 Connecticut Ave., N.W.,
Washington, D.C. 20036
(202) 296-5800

American Aerospace & Military Museum, Inc.
P.O. Box 1051,
Pomona, CA 91766
(714) 629-8310

American Aviation Historical Society (AAHS)
P.O. Box 996,
Ojai, CA 93023

American Bonanza Society
Chemung County Airport,
Horseheads, NY 14845
(607) 739-5515

American Helicopter Society
30 E. 42nd St., Suite 1408,
New York, NY 10017
(212)697-5168

American Institute of Aeronautics & Astronautics (AIAA)
1290 Avenue of the Americas,
New York, NY 10019

Antique Airplane Association
P.O. Box H,
Ottumwa, IA 52501
(515) 938-2773

Association of Aviation Psychologists (AAP)
Naval Safety Center, Code 1157,
NAS Norfolk, VA 23511

Aviation Distributors & Manufacturers Association
1900 Arch Street,
Philadelphia, PA 19103
(215) 564-3484

Aviation Hall of Fame, Inc.
Sheraton-Dayton Hotel,
Dayton, OH 45402
(513) 224-9601

Aviation Maintenance Foundation
P.O. Box 739,
Basin, WY 82410
(307) 568-2413 or 2414

Aviation/Space Writers Assoc.
Cliffwood Road,
Chester, NJ 07930
(201) 879-5667

Civil Aviation Medical Assoc.
141 N. Meramec Ave., Suite 4,
Clayton, MO 63105
(314) 862-1122

Experimental Aircraft Assoc.
P.O. Box 229,
Hales Corners, WI 53130
(414) 425-4860 or 4871

First Flight Society
P. O. Box 1903,
Kitty Hawk, NC 27949
(919) 473-2046

Flight Safety Foundation
1800 N. Kent Street,
Arlington, VA 22209
(703) 582-4100

Flying Chiropractors Assn.
528 Franklin Street,
Johnstown, PA 15905
(814) 536-6946

Flying Dentists Association
120-½ N. 5th Street,
Sleepy Eye, MN 56085

Flying Funeral Directors of America
678 S. Snelling Avenue
St. Paul, MN 55116
(612) 689-0895

Flying Physicians Association
801 Green Bay Road,
Lake Bluff, IL 60044
(312) 234-6330

General Aviation Manufacturers Association (GAMA)
Suite 1215
1025 Connecticut Avenue, N.W.,
Washington, D.C. 20036
(202) 296-8848

Helicopter Assoc. of America
Hangar D,
Westchester County Airport
White Plains, NY 10604
(914) 948-0614

International Flying Bankers Association
Armour Court
801 Green Bay Road,
Lake Bluff, Il 60044

International Flying Farmers
Municipal Airport
Wichita, KS 67209
(316) 943-4234

National Aero Club
3861 Research Park Drive,
Research Park
Ann Arbor, MI 48103

National Aeronautic Assn.
Suite 610, Shoreham Bldg.
806-15th Street, N.W.,
Washington, D.C. 20005
(202) 347-2808

National Aerospace Education Council
Suite 310, Shoreham Bldg.
605-15th Street, N.W.,
Washington, D.C. 20005

National Association of State Aviation Officials
Suite 802
1000 Vermont Ave., N.W.,
Washington, D.C. 20005
(202) 783-0588

National Aviation Trades Assn.
1156 15th Street, N.W.,
Washington, D.C. 20006
(202) 833-8210

National Business Aircraft Assoc.
Suite 401, Pennsylvania Bldg.
425 13th Street, N.W.,
Washington, D.C. 20006
(202) 738-9000

National Intercollegiate Flying Association (NIFA)
Parks College,
St. Louis University Parks Airport
Cahokia, IL 62206

National Pilots Association (NPA)
806 15th Street, N.W.,
Washington, D.C. 20005
(202) 737-0773

Ninety-Nines
P.O. Box 59964,
Oklahoma City, OK 73159
(405) 685-7969

OX-5 Aviation Pioneers
419 Plaza Bldg.,
Pittsburgh, PA 15219

Pilots International Association
2469 Park Avenue,
Minneapolis, MN 55407
(612) 546-4075

Professional Air Traffic Controllers Organization (PATCO)
2100 M Street, N.W.,
Washington, D.C. 20006
(202) 296-6443 and 6444

Soaring Society of America
P.O. Box 66071,
Los Angeles, CA 90066
(213) 390-4449

Society of Experimental Test Pilots
44814 N. Elm Ave.,
Lancaster, CA 93534
(805) 942-9574

University Aviation Association
Parks College
St. Louis University, Parks Airport
Cahokia, IL 62206

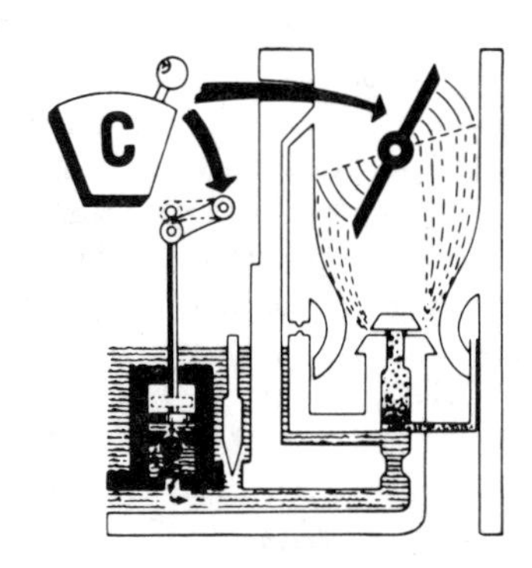

FAA General Aviation District Offices

Alaskan Region
1714 E. 5th Ave., Anchorage, AK 99501
5640 Airport Way, Fairbanks, AK 99701
Star Route 1, Box 592, Juneau, AK 99801

Central Region
228 Administration Bldg., Municipal Airport, Des Moines, IA 50321
Fairfax Municipal Airport, Kansas City, KS 66115
Municipal Airport, Wichita, KS 67209
9275 Glenaire Dr., Berkeley, MO 63134
Municipal Airport, Lincoln, NE 68524

Eastern Region
National Airport, Washington, D.C. 20001
Friendship International Airport, Baltimore, MD 21240
510 Industrial Way, Teterboro, NJ 07608
County Airport, Albany, NY 12211
Republic Airport, Farmingdale, NY 11735
Monroe County Airport, Rochester, NY 14517
Allentown-Bethlehem-Easton Airport, Allentown, PA 18103
Harrisburg-York State Airport, New Cumberland, PA 17070
North Philadelphia Airport, PA 19114
Allegheny County Airport, West Mifflin, PA 15122
Aero Industries, 2nd Floor, Sandston, VA 23150
Kanawha County Airport, Charleston, WV 25311

Great Lakes Region
P.O. Box H, Dupage County Airport, West Chicago, IL 60185
R.R. 2, Box 3, Springfield, IL 62705
St. Joseph County Airport, South Bend, IN 46628
5500 44th St., S.E., Grand Rapids, MI 49508
6201 4th Ave., S. Minneapolis MN 55450
4242 Airport Rd., Cincinnati, OH 45226
4393 East 175th Ave., Columbus, OH 43219
General Mitchell Field, Milwaukee, WI 53207

New England Region
1001 Westbrook St., Portland ME 14102
Municipal Airport, Box 280, Norwood, MA 02062
P.O. Box 544, Westfield, MA 01085

Northwest Region
3113 Airport Way, Boise, ID 83705
5401 NE Marine Dr., Portland OR 97218
FAA Bldg., Boeing Field, Seattle, WA 98108
P.O. Box 247, Parkwater Station, Spokane, WA 99211

Pacific Region
Rm. 715, Terminal Bldg., International Airport, Honolulu, HI 96819

Rocky Mountain Region
Jefferson County Airport, Broomsfield, CO 80020
Logan Field, Billings, MT 59101
P.O. Box 1167, Helena, MT 59601
P.O. Box 2128, Fargo, ND 58102
R.R. 2, Box 633B, Rapid City, SD 57701
116 North 23rd West Street, Salt Lake City, UT 84116
Air Terminal, Casper, WY 82601
P.O. Box 2166, Cheyenne, WY 82001

Southern Region
6500 43rd Avenue North, Birmingham, AL 35206
P.O. Box 38665, Jacksonville, FL 38665
P.O. Box 365, Opa Locka, FL 33054
Clearwater International Airport, St. Petersburg, FL 33732

Fulton County Airport, Atlanta, GA 30336
Bowman Field, Louisville, KY 40205
P.O. Box 5855, Jackson, MS 39208
Municipal Airport, Charlotte, NC 28208
P.O. Box 1858, Raleigh, NC 27602
Box 200, Metropolitan Airport, West Columbia, SC 29169
P.O. Box 30050, Memphis, TN 38103
Metropolitan Airport, Nashville, TN 37217

Southwest Region
Adams Field, Little Rock, AR 72202
Lakefront Airport, New Orleans, LA 70126
Downtown Airport, Shreveport, LA 71107
P.O. Box 9045, Sunport Station, Albuquerque, NM 87119
Wiley Post Airport, Bethany, OK 72008
International Airport, Tulsa, OK 74115
Redbird Airport, Dallas, TX 75232
6795 Convair Rd., El Paso, TX 79925
P.O. Box 1689, Meacham Field, Fort Worth, TX 76016
8345 Telephone Rd., Houston, TX 77017
P.O. Box 5247, Lubbock, TX 79417
1115 Paul Wilkins Rd., San Antonio, TX 78216

Western Region
2800 Sky Harbor Blvd., Phoenix, AZ 85034
Air Terminal, Fresno, CA 93727
2815 East Spring St., Long Beach, CA 90806
P.O. Box 2397, Oakland, CA 94614
International Airport, Ontario, CA 91761
Municipal Airport, Sacramento, CA 95822
3750 John J. Montgomery Dr., San Diego, CA 92123
1887 Airport Blvd., San Jose, CA 95110
3200 Airport Ave., Santa Monica, CA 90405
7120 Havenhurst Ave., Van Nuys, CA 91406
5700-C South Haven, Las Vegas, NV 89109
2601 East Plum Lane, Reno, NV 89502

The FAA Aeronautical Center address is P.O. Box 25082, Oklahoma City, OK 73125.

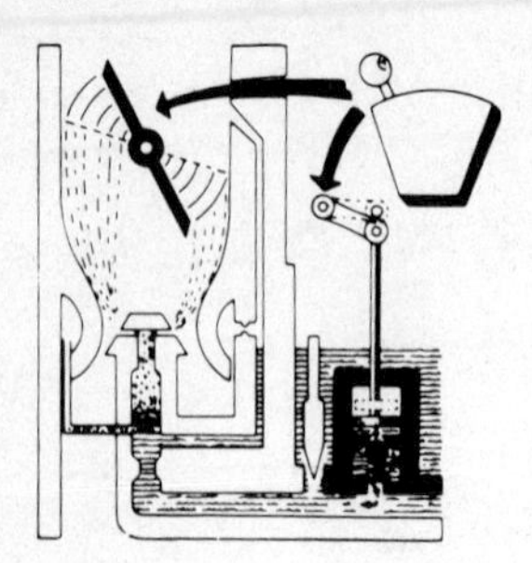

Index